Power Systems

More information about this series at http://www.springer.com/series/4622

Kaveh Niayesh · Magne Runde

Power Switching Components

Theory, Applications and Future Trends

 Springer

Kaveh Niayesh
Norwegian University of Science and
 Technology
Trondheim
Norway

Magne Runde
SINTEF Energy Research
Trondheim
Norway

ISSN 1612-1287
Power Systems
ISBN 978-3-319-84656-9
DOI 10.1007/978-3-319-51460-4

ISSN 1860-4676 (electronic)

ISBN 978-3-319-51460-4 (eBook)

Printed on acid-free paper

This Springer imprint is published by Springer Nature
The registered company is Springer International Publishing AG
The registered company address is: Gewerbestrasse 11, 6330 Cham, Switzerland

Preface

This book on power switching components is the result of two courses, which have been taught by the authors at the University of Tehran and the Norwegian University of Science and Technology for more than 10 years. It is designed primarily to serve as a textbook for a one-semester university course for graduate or senior undergraduate students in electric power engineering or other related disciplines. It may also be used for self-study purposes, for persons with a general background in electrical engineering, as it contains many detailed and self-explaining examples. A selection of some parts of this book may also be used to cover the important topics on current interruption as a part of a more general course in high voltage apparatus or high voltage technology.

The authors have made efforts to make the present book reasonably compact, but at the same time to cover all the important topics related to power switchgears that use mechanically separating contacts and interrupt current by extinguishing the electric arc that burns between the separating contacts. The book begins with a phenomenological description of the current interruption process, where the role of switching arcs is explained. Thereafter, the main features and related physical phenomena of different interrupting media are discussed in detail. In the application chapter that follows, three basic aspects are considered: the characteristics of the different switching duties, derivation of mathematical expressions and formulas describing the stresses the switching device are exposed to, and the testing methods used to qualify switching equipment. The switching technologies currently applied are then covered, in part by describing design and operation principles of typical devices. Then follows a short review of reliability and service experience of switching devices, together with a review of diagnostic methods being applied for checking the condition of switching equipment in service. The book ends with an outlook on certain development trends and challenges for the future generations of power switching devices.

This book is indebted to all the invaluable well-written preceding books in the wide field of power switching devices. Among these are "High voltage circuit breakers" by R. Garzon, "Switching in electrical transmission and distribution systems" by R. Smeets and his colleagues and "The vacuum interrupter" by P. Slade.

The first author would like to thank his previous colleagues and employers, the University of Tehran, AREVA T&D as well as ABB, also for giving the permission to use some photos in this book. He is grateful to all graduate students of the University of Tehran, who attended the course on the theory and applications of power switching devices and contributed to refining the contents of this course. The first author would also like to thank Prof. Hossein Mohseni, a great friend, teacher and colleague, and Prof. Klaus Möller, who introduced him to the exciting world of current interruption.

The second author wants to express his great appreciation to the late Prof. Jarle Sletbak, who introduced him to the field of current interruption and switchgear technology, and also to students, colleagues and other associates in academia and industry that over the years have contributed to making this a rewarding area to work in.

It was not possible to write this book without the backing of our current employers, the Norwegian University of Science and Technology (NTNU) and SINTEF Energy Research, both located in Trondheim, Norway. We appreciate this support. Last, but certainly not least, we gratefully acknowledge the loving support of our wives, Nazanin Arab and Ingeborg Bordal.

Trondheim, Norway   Kaveh Niayesh
November 2016   Magne Runde

Contents

Chapter 1
Introduction

Switching devices are inevitable in any electrical system, as they control power flow to different subsystems or components. From this perspective, their impedances have to change from almost zero (closed position) to nearly infinity (open position). A wide variety of methods, designs and constructions can be used to realize this function, depending on the desired current and voltage range.

The most commonly used methods to switch currents are by using mechanically moving contacts, and by means of semiconductor devices. Semiconductor-based switching components are the most favourable solution for many applications, but not in high current and high voltage systems. This is mainly due to the high losses of semiconductor switching devices in their on-state (when the switch is in "closed" position). This disadvantage combined with rather limited voltage ratings of available semiconductor switching components, cause semiconductor-based solutions for high current and high voltage applications to become expensive.

Mechanical switching devices, in contrast, can be designed in such a way that they have low resistances and thus low losses in closed position, typical several orders of magnitude lower than semiconductor-based switching devices.

By separating a current-carrying contact, a switching arc ignites and burns between the open contacts. The current continues to flow from one contact member to the other through this arc. The arc has a temperature of many thousand degrees, and at such high temperatures it is a good electrical conductor. In case of a successful current interruption of an alternating current, the conductance of this switching arc changes rapidly near current zero, from very high to very low values. The arc quenches, and this process enables the current interruption. Thus, in switching devices based on mechanically moving contacts, the switching arc plays a central role.

The focus of the present textbook is on the power switching components suitable for high current and high voltage applications in power transmission and distribution networks. Therefore, the primary interest is here on the mechanical switching devices, where the switching arc is used to perform the current interruption. The switching duties include planned connecting and disconnecting

© Springer International Publishing AG 2017
K. Niayesh and M. Runde, *Power Switching Components*, Power Systems,
DOI 10.1007/978-3-319-51460-4_1

different parts of the network (load current switching) as well as unplanned fault current switching. Hence, the switching devices are dealing with a wide current range, from a few amperes to short-circuit currents of several tens of kilo amperes and more. The present textbook is intended to give an in-depth insight to the theory and application of all parts of the devices that cover all necessary switching functions. It has been conceived as an appropriate resource for a graduate course on power switching components and can be used in any course on high voltage equipment. For this purpose, many worked examples and end chapter exercises on different aspects of switching devices in power networks are included.

1.1   General Aspects of Switching in Power Grids

The switching tasks or duties in a power system can be grouped into two main categories: load current and fault current switching. Load current switching is typically a planned change in the grid configuration, e.g. energizing or de-energizing a power cable or an overhead line, or starting or stopping an electric motor. Thus, load switching is in all aspects a normal part of the day-to-day operation of power components and power networks. When disregarding initial transients that in some cases occur, the currents to be switched are not higher than the maximum rated load current for the component at the actual location of the system.

Equally important as load current switching is the much rarer and but far more demanding fault current switching. Faults in the electric power grids are unavoidable, because power systems and equipment are not technically perfect solutions, but only trade-offs between meeting minimum technical requirements and reducing the manufacturing cost. In other words, even though it is technically possible to reduce the fault probability to almost zero, e.g. by over-dimensioning of the equipment, it will not end up as an optimal solution from a technical-economical point of view.

In many faults in power grids, e.g. an electrical insulation failure occurring due to lack of overvoltage withstand capability under a lightning strike, a low impedance path through an electric arc is created between two electrically conducting parts at different voltage levels. This event is called a short circuit in the network. With the occurrence of a short circuit, very high currents are flowing through the fault location. To limit the damages caused by the arc and the fault current to the different components, and to keep the network operating stably, the fault has to be cleared quickly. In practice, the short circuit current has to be interrupted as fast as possible. To accomplish fault detection at early stages, protection systems including measurement devices such as current transformers (CT) and voltage transformers (VT) as well as data processing devices have been developed. By desirable performance of power protection systems during a fault, an opening command is sent to the appropriate switching device in the network. The function of the switching component is then to interrupt the current and thereby to clear the fault. From this perspective, it can be realized that the power switching devices play a decisive role in the power network protection; therefore, their reliable performance is of major interest for network designers and operators.

From the perspective of producers and consumers of electrical energy, it is not desirable if the energy supply is impaired in the whole grid because of any fault occurring somewhere in the network. For this purpose, it is necessary to have a significant number of switches along the path from the energy producer to the energy consumer. This makes it possible to selectively isolate different parts of the grid in case of a fault, without disturbing the energy supply to the other consumers. That is the reason why at all voltage levels in the network and in all physical locations, such as near the generator, in transmission and distribution substations as well as near the end consumers, different types of power switches are employed. This is shown schematically in Fig. 1.1, where power switches are depicted in the form of solid rectangles. It is clear that the switching components used in different positions are exposed to different stresses and therefore have different capabilities and features.

From the above it is clear that power switching devices are handling large electric energies. The current flowing through the arc that is created when separating the contacts of a switch results in a significant energy dissipation in the switch. The incurred energy losses in turn lead to a higher arc temperature and an increased electrical conductivity of the arc, so that interrupting the current cannot be realized only by mechanically separating the contacts from each other. For instance, a one kilo-ampere electric arc burning in an air gap of two meters remains stable in the sense that it does not extinguish by itself.

Consequently, controlling the dissipated energy in the switching arc as well as including design measures to decrease the electrical conductivity by cooling the arc to such an extent that it eventually quenches, causing power switching components to become complicated constructions. Obviously, understanding the behaviour and properties of the switching arc is crucial in this context.

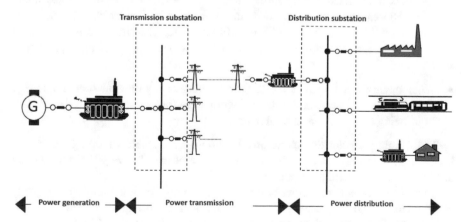

Fig. 1.1 Schematics of a typical power system including generation, transmission and distribution with the emphasis on the position of power switching devices

1.2   Switchgear Requirements

Irrespective of the area of application, every switching component must fulfil the following requirements:

- When closed, it must be an almost perfect electrical conductor.

When the switch is in closed position, heat generated by Ohmic losses rises the temperature, so to obtain a high rated load current a low resistance is necessary. Furthermore, the switch has to be capable of carrying all possible currents that may flow through the network without opening unintentionally. In the event of a fault in the network, short circuit current may pass through a number of switches, but only one or a few of them should be opened. Consequently, currents that can pass through the switches are either high currents for a short time (e.g. short circuit current for a maximum of 1–3 s) or the much lower continuous load currents for long times.

- When open, it must be a perfect insulator.

In open position, there is no galvanic connection between the two sides of the switch. The switch must remain open regardless of what happens in the network. This implies that the switch needs to withstand all possible applied voltages, including transient overvoltages (e.g. lightning surges and overvoltages generated by switching operations in the network) as well as power frequency overvoltages and rated voltage. In this context, dielectric strength across the open contacts is of crucial importance.

- When closed, it must be able to interrupt any current up to its maximum rated breaking current at any time, without generating unacceptably large overvoltages.

The maximum rated interruptible current is one of the main differentiators between different types of power switching components. For a circuit breaker it is the short circuit current, in case of a load break switch it is the rated nominal load current, and in case of a disconnecting switch the current interrupting capability is almost zero.

- When open, it must be able to close the contacts at any time, including against a short circuit, without welding them together (as this would prevent the breaker from being able to open at a later occasion).

The number of closing operations it should be capable of doing during its lifetime, and the maximum required closing current are also very different among different power switching components.

These requirements are rather tough and demanding, and in some aspects partly contradictory. The emphasis put on the different requirements depends on the application. Different application cases are presented in detail in Chap. 3.

Table 1.1 Maximum current values for some important switching duties

Switching duty	Current	
	$U \leq 36$ kV	36 kV $< U \leq 1100$ kV
Interruption of load currents/fault currents in distribution and industry networks	<2000 A/<90 kA	
Interruption of load currents/fault currents in transmission grids		<4000 A/<90 kA
Interruption of load currents/fault currents in generator breakers	<25,000 A/< 300 kA	
Out of phase interruption		<20 kA
Interruption of small inductive currents		
No-load transformer	<400 A	<20 A
Reactor	<1250 A	<400 A
Motor	<2000 A	
Interruption of capacitive currents		
No-load overhead line		<400 A
No-load cable	<100 A	
Capacitor bank	<400 A	<400 A
Closing against short circuit	<150 kA	<75 kA

Nowadays, switching equipment for voltages up to 800 and 1100 kV are in normal use. Some of these have to be able to interrupt short circuit currents approaching 100 kA. Generator circuit breakers operate at lower voltages, but the maximum currents they have to interrupt can be very large, up to several hundreds of kilo amperes.

The most important switching duties, with the associated maximum currents involved, are shown in Table 1.1. In most cases, the currents are significantly lower than these rather extreme values.

1.3   General Design of a Power Switching Device

Since switching devices operate by quickly separating or mating the current-carrying contacts, it is necessary to have a subcomponent for creating the mechanical movement. In this subcomponent, known as the *operating mechanism* or the *drive mechanism*, mechanical energy is stored and transferred to the contacts of the switching device to move them with an appropriate speed in order to close or open them. Desirable interaction between drive mechanism and *interruption chamber*, where the contacts are located and the switching arc burns and is interrupted, ensures a successful current interruption.

In addition to the interruption chamber and the drive mechanism, there is a third subsystem, which handles fault detection and gives the opening and closing

Fig. 1.2 General design of a
power switching device

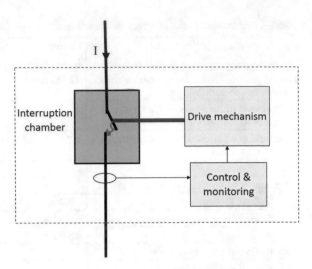

commands to the drive mechanism. This subsystem can be found locally in the
power switching device itself, or be a part of the control system of the network and
located outside of the switching device. The three important subcomponents of a
power switching device and their relationships are shown schematically in Fig. 1.2.

From the perspective of the high voltage part of the power system, only the
current carrying part of the power switch, namely, the interruption chamber is seen.
The other subcomponents of the switching device, which are electrically insulated
from the high voltage part of the power system, are indiscernible. So the other two
subcomponents (drive mechanism and control and monitoring circuitries) are only
considered in this book when they interact directly with the current interruption
process in the interruption chamber.

1.4   Stresses on Switching Equipment

The materials involved and the various subcomponents of a switching device are
exposed to several types of stresses during switching operations, and when being in
open or closed position. Hence, when designing switching devices a large number
of factors and stresses must be taken into account.

1.4.1   Mechanical Stresses

It takes at least a few power cycles to clear a short circuit fault in a grid, i.e to open
the circuit breaker and electrically separate/isolate the failed components from the
rest of the system. During this period, the switchgear and other components are

exposed to mechanical stresses generated by the short circuit current. The electromagnetic or Lorentz force F is given (per volume) as the vector product of current density J and magnetic flux density B: $F = J \times B$. As the flux density is proportional to the current, the Lorentz forces between two electrical conductors are proportional to the current squared and inversely proportional to the distance between the conductors. The forces are also highly dependent on the shape of the current paths.

Other mechanical stresses occur when the contacts open during an interruption. The stresses are typically caused by the increased pressure in the interruption chamber, in particular when the arc extinguishing medium is oil. In this case, the arc evaporates the oil, and this causes a substantial pressure rise, increasing with amperage.

Furthermore, since modern switchgears usually rely on very fast moving of the contacts during interruption, considerable masses are subjected to heavy accelerations and forces in the beginning and towards the end of a switching operation. The mechanical design of the moving parts of the switchgear is mainly determined by these large stresses of the driving mechanism, including the dashpot.

1.4.2   Thermal Stresses

Even though short circuit currents lead to a short term heating, the continuous rated current is the main cause for thermal stresses in switching devices. The energy dissipated due to the electric resistance of contacts, joints and terminations in the primary circuit is the prime heat source. The magnitude of the contact resistance depends on the contact material, the design and the contact pressure. It is also crucial that heat is carried away efficiently to avoid unacceptably high temperatures. These matters are closely linked to the design and may differ greatly among different breaker technologies. Moreover, heat may also be generated by eddy currents or by hysteretic losses in ferromagnetic materials.

Parts inside the interrupting chamber are exposed to thermal stresses caused by the electric arc. This may lead to contact erosion as well as wear on nozzles and other parts in intimate contact with the extremely hot arc.

1.4.3   Dielectric Stresses

Since it is required that power switches should behave as perfect insulators when open, great demands are put on both the interrupting and the insulating media. The choice of interrupting medium is particularly complicated as it also serves as the insulating medium when the contacts are in open position. The electrostatic field distribution must be taken into consideration when designing insulation system and contacts. However, usually most of the design efforts are concentrating on obtaining a high dielectric strength immediately after the arc is quenched, while the contacts are still moving. In most cases, the distance between the contacts is still not at its maximum at the instant of current interruption, and the insulating medium has a reduced dielectric strength due to presence of residues in the switching gap generated by the recently extinguished electric arc.

1.5   Types of Switching Components in Power Networks

All switching devices in power networks have to fulfil the first two requirements listed in Sect. 1.2. The last two requirements can be used to categorize switching devices into different types according to the switching duties they should be able to perform:

The *disconnector switch* is able to open a circuit when it is energized, but not to interrupt any load or short circuit currents. Hence, the interrupting capability is limited to very small capacitive charging currents. Normally, for safety reasons it is required that the open circuit in the disconnector switch is visible.

The *load break switch* is able to interrupt currents less or equal to the rated load current, i.e. the maximum continuous current the system is rated for. Generally, there are restrictions on the cos ϕ of the current; typically cos $\phi \geq 0.7$ for currents that the load break switch should be able to interrupt. This limits the stresses on the switch. The load break and disconnector switch functions are often combined into one device, referred to as a *switch-disconnector or disconnector-switch*. Fuses are normally used in series with such a device, for interrupting short circuit currents.

The *circuit breaker* has the most demanding task, as it should be able to interrupt all currents, including short circuit currents. This implies that it is subjected to the largest current and voltage stresses. Circuit breakers are very important for clearing faults in a grid, particularly for voltages greater than about 36 kV where fuses cannot be used as protective devices for short circuits. Moreover, the circuit breaker has a safety function in transmission grids, as it protects large equipment installations and is supposed to limit the damage in case of a short circuit. This means that the *reliability* of the circuit breaker becomes very important.

Earthing switches are used to ground parts of a grid. Two types exist. *High-speed earthing switches* quickly connect energized parts to earth. These must be able to carry large currents (to earth), but do not interrupt currents. There are also requirements for how rapid the operation should be. To ground components or lines that are already de-energized, slower operating earthing switches are used. Typically, this is required before maintenance or other types of works can be initiated. Earthing switches are often constructed in the same way as disconnecting switches, but instead of linking together different parts of the primary circuit (i.e., the high voltage circuit), they connect the primary circuit to ground.

The general terminology used in the field of power switching devices is not always very concise. The power engineering community uses the terms "switching device", "switchgear", "breaker", "switch", and "interrupter" to a large extent as synonyms. In addition, "switchgear" is used both for a single switching component and for an assembly of many switching components. The definitions above of specific types of power switching devices according to their duties are however, well established and clear.

Another criterion for the classification of switching devices is their rated voltage. Although all power components with a rated voltage of more than 1000 V are considered, according to IEC standards, as high-voltage power equipment, another definition for voltage levels from the perspective of manufacturers and users exists. Voltages up to 1000 V are considered *low voltage*, a rated voltage in the range of 1000 V to a few tens of kilovolts (for example, the IEC voltage classes of 6, 12, 24, 36, 52 and 72.5 kV) as *medium voltage (MV)* and higher rated voltages as *high voltage (HV)*. The reason for this second and quite commonly used classification is that switching technologies may differ greatly between these three voltage ranges.

The terms *distribution level voltage* and *transmission level voltage* are also quite common, typically referring the 6–36 kV, and 145 kV and above, respectively. Sometimes the term *sub-transmission level voltage* is applied, meaning the range from 52 kV and up to and including 145 kV.

The complexity of switching devices increases with increasing current interrupting capability and rated voltage. For the lowest current and voltage ratings switching devices are realized with the arc burning in an air environment and without any complicated arc quenching arrangements. Interrupting larger currents and serving at higher voltages require more sophisticated designs and also arcing media better suited for the purpose. In old generations of circuit breaker designs, high pressure air blast or oil were used in the interrupting chamber to achieve sufficient ratings, see Fig. 1.3.

In the last decades, current interruption technology in vacuum and SF_6 gas has been significantly developed, and switching devices utilizing these two technologies gradually replaced oil and compressed air circuit breaker technologies.

Fig. 1.3 Insulating mediums of old generation power switching devices for different rated voltages and maximum interruptible currents

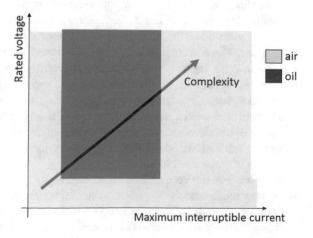

Fig. 1.4 Insulating mediums of power switching devices for different rated voltages and maximum interruptible currents (present status)

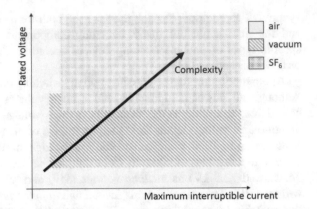

Currently, vacuum and SF_6 technologies dominate at the medium voltage and high voltage levels, respectively, as shown in Fig. 1.4.

Although SF_6 technology in the medium voltage range is cost competitive to vacuum technology, sales of SF_6 based medium voltage switching devices are recently going down worldwide and the market share of vacuum-based technology is increasing. This is mainly because there is a lot of pressure from policy makers not to use SF_6, wherever a technical-economic alternative exists, as this gas possesses a very strong greenhouse impact (One kilogram of SF_6 has an equivalent environmental impact as more than 20 tons of carbon dioxide).

Basic working principle of all technologies referred to in Figs. 1.3 and 1.4, along with their features and specifications are discussed in detail in Chap. 4.

1.6 Outline of the Book

Considering the crucial role of the electric arc generated between contacts of switching devices during the current interruption process, the second chapter of this book is dedicated to the basic theory of current interruption and important related parameters. The current flowing through the switching device as well as the voltage across the open contacts just after current interruption are identified as significant parameters impacting the behaviour of the switching arc. These are the main parameters varying from one application to the other. Various approaches to understand and model the switching arc as a key element in interruption of the current will be addressed. A simple approach to describe the interaction between the switching arc and the power network, namely the so-called "black box modelling" approach is considered in detail.

Chapter 3 is the core of this book, especially for readers interested in applications and requirements of power switching components. The idea is to present various requirements, which are supposed to be fulfilled by switchgears in distribution and transmission networks, to highlight special features of switching devices making them able to cope with the related stresses, and finally to show how these features may be evaluated based on appropriate testing methods. This includes switching of load currents (resistive, inductive as well as capacitive) and switching of fault currents (i.e., short circuit currents) in different network configurations.

Different technologies used in switchgears are discussed in Chap. 4. These include vacuum, gas, oil and magnetic air circuit breaker technologies.

In Chap. 5, reliability aspects of operation of power switching devices as well as their possible fault mechanisms are discussed based on service experiences. Different condition assessment and diagnostic methods are reviewed.

Chapter 6 is devoted to presentations of technological trends aiming at further developing power switching components. A detailed treatment of this topic is outside the scope of this book, but as it is important to make the graduate students familiar with evolutions and ongoing developments of power networks, a few topics are briefly addressed. These include current interruption in Direct Current (DC) networks, limitation of fault currents and SF_6-free high voltage power switching devices.

Chapter 2
Current Interruption Basics

In this chapter, the principle of current interruption in power switching devices with mechanically separating contacts is presented. The interruption is associated with initiation and extinction of a switching arc. First, a qualitative description of current interruption in power networks with various load types is given, and important parameters and concepts are introduced.

In the second part, the switching arc as the key element in current interruption in mechanically opening switches is considered in detail. Relevant physical phenomena of the switching arcs in different types of interrupting media, as well as methods for arc modelling are discussed comprehensively.

2.1 A Phenomenological Description of Current Interruption in an AC Power System

2.1.1 Contact Separation and Switching Arc

A switching device contains one or more pairs of contacts in each phase. Under normal service, these are in closed position and current passes through the switch. When an opening command signal is sent to the switchgear, its driving mechanism will set the contacts in motion so that they start moving away from each other. As explained earlier, the current is not interrupted at the time of mechanical separation of the contacts, but continues to flow through an electric arc that ignites in the gap between the opening contacts. The electric arc consists of a mixture of electrons, neutral particles, and positive and negative ions. The temperature of the arc is very high due to the energy dissipated in the arc by the current flow, making it a reasonably good electrical conductor. The key task of a power switch is to control the energy losses during arcing, as well as to provide appropriate measures to make the arc unstable and cause the current to be interrupted near its zero crossing.

© Springer International Publishing AG 2017
K. Niayesh and M. Runde, *Power Switching Components*, Power Systems,
DOI 10.1007/978-3-319-51460-4_2

The voltage drop along the arc through most of a current half cycle is approximately constant and much lower than the rated voltage of the power network. As the current passes zero, the input power to the arc also becomes zero, and the processes responsible for generation of electrical charge carriers in the arc cease. The working principle of a switching device is essentially to get rid of the available charge carriers between its contacts so efficiently that the gap becomes virtually insulating as the current reaches its zero crossing, and then to quench the arc and to interrupt the current at this instant. In case of switching arcs in gaseous interrupting medium, this process is associated with the cooling of the electric arc.

After the arc has been quenched at current zero, a voltage generated by the surrounding power network arises across the contact gap of the switching device. This is called the *recovery voltage*, and its amplitude and steepness determine whether a new arc ignites after current zero, that is, whether or not the interruption has been successful.

A simplified single-phase circuit diagram where a short circuit has occurred close to the breaker is shown in Fig. 2.1. Figure 2.2 shows the current through the

Fig. 2.1 Single-phase system, short circuit close to the breaker

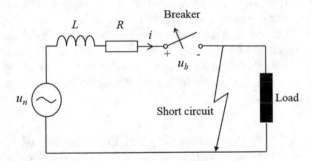

Fig. 2.2 Current and voltage during a current interruption

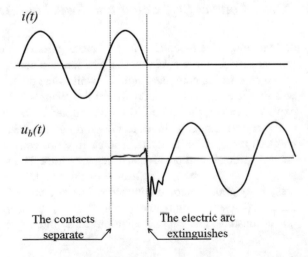

breaker and the voltage drop across the contacts during the interruption process as described above.

The following factors play a major role in the success or failure of an attempted current interruption:

- **Arc current**: The larger the amplitude of the current, the higher becomes the arc temperature and the density of electric charge carriers generated in the arc. This makes it much more difficult for the contact gap to become insulating after current interruption at current zero. In addition to the current amplitude, the current steepness (*di/dt*) near current zero is of crucial importance for the current interruption process. A higher *di/dt* means that the switching gap has less time to change from conducting to insulating state, before the gap is exposed to the transient recovery voltage.
- **Arcing time**: Since there is no control on the mechanical opening and closing moment of most switching devices, the instant when the contacts of a switching device separate is a random variable. If the contacts separate just before the current zero crossing, the current will not be immediately interrupted but continues to flow until the next current zero crossing. Therefore, the arc duration is typically in the range of 0.5–1.5 times the length of a current half cycle. By longer arc durations, the energy dissipation in the breaker increases, and interruption of the current usually becomes more difficult.
- **Arc voltage**: The input power required for an arc to remain stable depends on the medium in which the arc is burning. This defines the arc voltage that is observed. The arc voltage is dependent on the design and the materials—including the type of interrupting medium—of the switching device itself and not on the rated voltage of the power grid.
- **Transient recovery voltage (TRV)**: After current zero crossing, in case of a successful interruption, a voltage is generated from the network onto the terminals of the switching device as the result of energy oscillations between the energy storage elements of the network. This is called the recovery voltage, and it normally has an initial transient part, the Transient Recovery Voltage (TRV), This voltage can accelerate the remaining electric charge carriers present in the contact gap, increasing the chance of getting charge carrier multiplication by impact ionization and finally lead to a breakdown of the switching gap and formation of a new arc. Thus, the TRV is one of the key factors determining a switching device ability to interrupt a current. In particular, the *Rate of Rise of the Recovery Voltage (RRRV)* is of great importance. The TRV and short-circuit current are determined by the network where the switching device is installed, and differ for different switching applications, see Chap. 3.

The switching arc should have a very high electrical conductance, so that the current can flow through it from the moment of contact separation to current zero, without dissipating excessive amounts of power in the interruption chamber. After current zero, its conductance has to go rapidly to zero, so that no current will flow through the arc causing the current to be interrupted. The total energy dissipation during arcing in a switching device E_{loss} can be expressed as follows:

$$E_{loss} = \int_{t_{sep}}^{t_{cz}} u_{arc} \cdot i_{arc} \cdot dt \qquad (2.1)$$

where t_{sep} is the moment of contact separation and t_{cz} the current zero crossing. u_{arc} and i_{arc} are arc voltage and arc current, respectively.

Consequently, the key processes for achieving a successful current interruption are to control the dissipated energy in interruption chamber during arcing as well as to rapidly decrease the arc conductance from very high values to near zero at current zero crossing.

The breaker is said to *re-ignite* or *re-strike* if a new arc is formed after a current zero crossing. A re-ignition occurs immediately after current zero, while a re-strike is defined to occur at least a quarter of a power cycle later. Re-ignitions are divided into two categories: thermal and dielectric re-ignitions.

The temperature of the electric arc channel is still high as the current passes through zero, and thus some electrical conductivity remains. When the recovery voltage then builds up, some power dissipation takes place in the arc path. If the cooling is efficient, the temperature nevertheless drops, conductivity reduces, and the current goes towards zero. However, the cooling might not be sufficient, and the temperature and conductivity may then rise and a new electric arc is formed. This is referred to as a *thermal re-ignition* as it is caused by a thermal instability in the electric arc. The temperature in the contact gap is closely correlated to the amplitude of the current that is being interrupted. Thermal re-strike occurs immediately (up to a few microseconds) after current zero and is greatly dependent on the recovery voltage shape, especially its steepness, during this period.

If a thermal re-ignition is avoided, the voltage across the contacts increases. Even if there is practically no electric conductivity left in the contact gap, this area is dielectrically stressed. A re-ignition will occur if the recovery voltage at any time exceeds the dielectric strength of the gap. This is referred to as a *dielectric re-ignition*.

The dielectric strength increases with time as the contact members move apart. However, as the dielectric strength of a gas is inversely proportional to its absolute temperature, the condition of the gas in the gap also plays a role. The gas may still be warm due to the electric arc that has been burning in the gap. Often it is found that the most critical time as to whether a dielectric re-strike will occur is a few milliseconds after current zero. Thus, the possibility of having a dielectric re-strike is influenced by the shape and amplitude of the recovery voltage in this period.

Interruption of an alternating current can therefore be seen firstly as a race between heat generation and cooling in the contact gap (risk of thermal re-ignition);

and then secondly as a race between voltage build-up and dielectric strength in the contact gap (risk of dielectric re-ignition).

2.1.2 Recovery Voltage

As mentioned above, the recovery voltage is determined by the properties of the power system in which the switching device is installed, in particular the type of load being interrupted. Idealised examples of recovery voltages in single-phase systems, after interruption of a resistive, capacitive and inductive load are shown in Fig. 2.3. The voltages u_l and u_r are the voltages at the left and right side of the breaker, respectively. The arc voltage is assumed negligible compared to the system voltage and the voltage across the switchgear before the arc extinguishes at current zero is $u_{breaker} = u_l - u_r = 0$.

Current and voltage are in phase when the load is resistive (Fig. 2.3a), and the voltage drop across the breaker follows the source voltage. In this case, the recovery voltage has no transient part and $u_{breaker}$ never becomes greater than the source voltage.

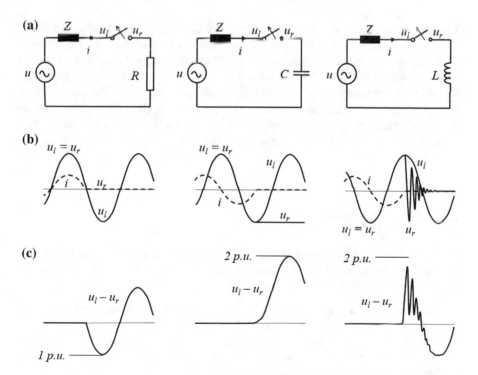

Fig. 2.3 Current and voltage waveforms (schematic) during interruptions of resistive (**a**), capacitive (**b**) and inductive (**c**) loads. The impedance on the source side of the breaker is ignored

When the load is purely capacitive (Fig. 2.3b), the current is interrupted when the source voltage is at its maximum. The voltage at the left side of the switchgear follows the source voltage, while u_r remains at its maximum value due to the capacitive charging at the load side. The maximum amplitude of the voltage across the contacts thus becomes twice the source voltage amplitude.

The idealised cases with resistive and capacitive loads give easy current interruption, as there are no voltage transients following the current zero crossing. Inductive load (Fig. 2.3c) in combination with stray capacitances (not included in the figure), on the other hand, gives an oscillatory circuit at the load side. Hence, u_r quickly goes to zero, but with a high frequency transient voltage component. If the damping in the circuit is low, the maximum amplitude of $u_{breaker}$ becomes twice the source voltage amplitude. Consequently, the chance for having a re-ignition is greater in this case, due to both the steepness and the amplitude of the TRV.

These simple and idealised examples serve as illustrations and an introduction to a more thorough treatment of different switching duties and the associated recovery voltages that will follow in subsequent chapters.

2.2   Switching Arcs

When gases or metal vapour are heated to very high temperatures a large part of the molecules decompose and break down into a mixture of atoms and other neutral particles, free electrons and positively and negatively charged ions. This mixture is called a *plasma* and is the main ingredient of the electric arc, which always is formed when the contacts of an energized switchgear separate. The properties of arcs burning in gases at atmospheric pressure and above (known as high pressure switching arcs) are treated in Sect. 2.3, while physical foundations of the arcs burning in very low pressure (vacuum) environments are discussed briefly in Sect. 2.4.

2.2.1   Arc Initiation

Sending an opening command signal to a power switching device causes its contacts to separate soon after and a *switching arc* is formed. Different stages of this process may be explained as follows:

– When the contact is closed current flows from one contact member to the other one, not through the entire apparent contact interface, but through a limited number of contact points (see Fig. 2.4a). The number and resistance of these minute conducting areas depend on the contact material and the force pressing the two metallic contact members together.

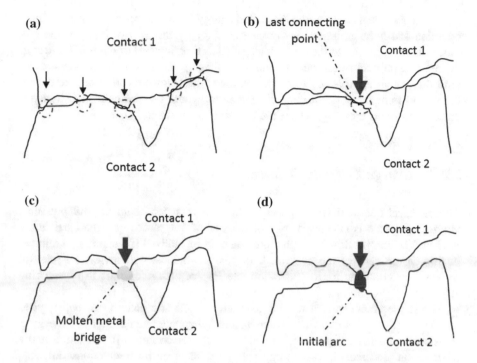

Fig. 2.4 Various stages of initial arc formation by separating current-carrying contacts **a** contacts in the closed position **b** contacts are connected only at the last point **c** a molten metallic bridge is formed at the last connecting point **d** the initial arc formed by evaporation and ionization of the molten metallic bridge (Note that the contacts surface roughness is exaggerated in this figure to better explain the concept)

- The very high current density in the last connecting point results in formation of a liquid metallic bridge between the contacts. By parting the contacts from each other, its cross-sectional area decreases and the current density further increases. Eventually, the liquid bridge evaporates and is ionized. This process leads to formation of the so-called *initial arc*.
- From now on, motion of charge carriers generated in the arc makes it possible to keep the current flowing from one contact to the other one. If the formation of the initial arc happens in the vicinity of an ionisable medium (like gas or oil in switching devices), this medium is also ionized. The arc characteristics are then determined by the material properties of the surrounding medium. If no ionisable medium surrounds the initial arc (like in vacuum switchgear), electric charge carriers can only be supplied from the contacts, and therefore, the arc characteristics are determined solely by the contact material.

Figure 2.4 shows the various stages of the arc initiation during separation of current-carrying contacts. After formation of an electric arc between the two contacts, the current flow will continue until the arc becomes unstable. The switching arc remains stable as long as the dissipated energy from the arc (e.g. in form of heat radiation and light emission) is compensated for by the energy input to the arc. Since the input energy to the arc is determined as time integral of the input power to the arc (the product of arc current and arc voltage), the arc usually remains stable until the current approaches zero. Hence, the current flows through the arc during the period from the time of contact separation, where the switching arc initiates, until current zero crossing, where the arc becomes unstable.

2.2.2   Charge Carriers Balance

The electrical conductivity of gases is under normal temperatures and pressures almost zero. This is because very low number of free electric charge carriers are present. The current flow through a gas can only be realized if a significant number of free charge carriers are generated. In this section, we will review all relevant mechanisms contributing to generation and loss of charge carriers in a switching arc.

The various mechanisms for the generation of electric charge carriers in gases are also referred to as *ionization mechanisms*. When sufficient amount of energy is transferred to a neutral particle, an electron may detach and a free electron and a positive ion are created. The energy may be transferred by interactions (impacts, collisions) with electrons, neutral atoms or positive ions, or may be acquired by absorbing radiation. It is also possible for electrons within the metal contacts of a switching device to overcome the energy barrier and be emitted the gas (*electron emission mechanisms*). These two types of generation mechanisms of charge carriers are shown schematically in Fig. 2.5.

The generated free electrons may recombine with the ions and this decreases the number of available charged particles. Another important process is the attachment of the electrons to atoms and formation of negative ions, which have by far much higher masses. During this process, very mobile charged particles are converted into charged particles with very low mobility, which do not significantly contribute any more to the current flow through the switching arc.

2.2.3   Ionisation Mechanisms

In ionisation processes due to impacts and collisions, it is common to distinguish between *thermal ionisation* caused by random impacts in a gas at high temperature, and *impact ionisation* caused by acceleration of the electrons in an electric field in the time interval between two impacts. In the first process, the energy is solely associated with the high temperature, whereas the energy in the last process comes from the electric field, at least partly.

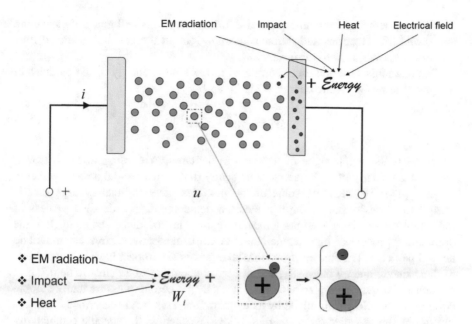

Fig. 2.5 Schematically shown different charge generation mechanisms in a switching arc including ionization and electron emission processes

2.2.3.1 Thermal Dissociation and Ionisation

Occurrence of arc in switching devices is associated with generation of heat. By increasing the temperature of the gas, the gas molecules begin to decompose to their building atoms. Decomposition or splitting of a molecule into two or more smaller neutral particles is referred to as dissociation, such as the dissociation of a nitrogen molecule into two nitrogen atoms ($N_2 \rightarrow 2N$).

Dissociation occurs when a gas is heated to temperatures where the kinetic energy of the particles exceeds the binding energy of the molecules. The dissociated fraction, x_d, of the gas molecules at a given temperature T and pressure p is [1]:

$$\frac{x_d^2}{1 - x_d^2}p = K_1 T^{3/2} e^{-\left(\frac{W_d}{kT}\right)} \tag{2.2}$$

where K_1 is a constant and W_d is the dissociation energy of the gas.

Dissociation is a far more complicated process for a gas molecule consisting of many atoms, e.g. SF_6, compared to a simple diatomic gas as nitrogen. The dissociation of SF_6 takes place through several steps where the molecule is gradually split into smaller fragments. Finally, when the temperature has reached a sufficiently high level a complete dissociation into fluorine and sulphur atoms occurs.

If the gas temperature continues to rise further, electrons will gradually leave the atoms and free electrons and different positive ions in the gas are generated (this process is known as *thermal ionisation*).

Thermal ionisation is closely correlated to the kinetic energy of the particles in the gas. The mean kinetic energy W_k is given by:

$$W_k = \frac{1}{2}m\overline{v^2} = \frac{3}{2}kT \tag{2.3}$$

where $\overline{v^2}$ is the mean value of the velocity squared. According to the Maxwell theory, the velocity of each particle statistically distributes around a mean velocity.

At a given temperature, some of the particles have sufficiently high kinetic energy to cause ionisation through electron detachment. The detailed process of ionisation depends on the gas temperature and is not discussed here, but the dominating process at high temperatures is collisions between free electrons and neutral particles. The free electrons that take part in the impact have acquired most of their kinetic energy thermally, and not by acceleration in an electric field. Thus, the density of generated charge carriers in a gas at a certain pressure under thermal equilibrium is solely determined by its temperature. This fact is described in plasma physics by the so-called *Saha equation* [2]. Consequently, the specific conductivity of any gas and its degree of ionization are strongly dependent on its temperature. The fraction of the particles in a gas that are thermally ionised are, similarly as for dissociation, given by:

$$\frac{x_i^2}{1 - x_i^2}p = K_2 T^{5/2}e^{-\left(\frac{W_i}{kT}\right)} \tag{2.4}$$

where K_2 is a constant, and W_i is the energy needed for single ionisation of a gas molecule. The ionisation energy is also often expressed as

$$W_i = eV_i \tag{2.5}$$

where e is the elementary charge, and V_i the ionisation potential. The variation of x_i due to variations in temperature and ionisation energy is shown in Fig. 2.6, and the ionisation energies for some particle types are listed in Table 2.1. It can be seen that at normal temperatures, the number of ionized atoms is very small, so that the gas has very low electrical conductivity. As the temperature increases above a certain level, the number of ions increases abruptly. By further increase of temperature, the gas becomes fully ionised.

The energy needed to detach one or two electrons is W_{i1} or W_{i2}, respectively. The table also contains the ionisation energies for single atoms. This is of interest since some gases dissociate before they are thermally ionised. Consequently, the ionisation potential for the dissociation products is more relevant than that of the complete gas molecule.

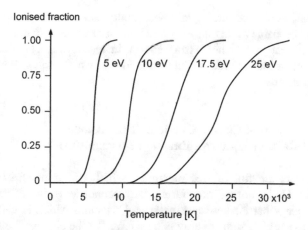

Fig. 2.6 The fraction ionised by thermal ionisation as a function of ionisation energy and temperature. The value of K_2 in (2.4) is set to 3.16×10^{-7}

Table 2.1 The thermal ionisation energy for some gases [3]

Particle type	W_{i1} (single ionisation) (eV)	W_{i2} (double ionisation) (eV)
N_2	15.6	
N	14.5	44.1
O_2	12.5	
SF_6	16.2	
S	10.4	33.8
F	17.4	

2.2.3.2 Impact Ionisation and Excitation

Collision or impact ionisation generally refers to ionisation due to an impact between a gas particle and an electron that has gained energy through acceleration in an electric field. This is not a precise definition as the kinetic energy obtained from the field adds to its thermal energy. The electron energy has to be equal to or greater than the ionisation energy of the particle it collides with in order for the impact to cause ionisation. As mentioned earlier, it is only just after current zero crossing, where the electric field is strong enough and temperature low enough for collision ionisation to be of any importance.

In gases and arcs with high pressure and/or low electric fields, the electron may experience many elastic collisions between each inelastic (which may cause ionisation). Due to these elastic impacts, the velocity of the electron along the field will have a randomly directed component, much in the same way as the thermal movement. In strong electric fields, it is said that the *electron temperature* is greater than the heavy particle/gas temperature, and it is thus difficult to separate between thermal and collision ionisation in an arc.

In principle, an ion accelerated in a field could also cause impact ionisation, but this is far less probable as an ion, in contrast to an electron with much lower mass, would lose a great part of its "extra" energy in elastic impacts. Therefore, very strong fields are needed in order to "heat" the ions to temperatures above the neutral particle temperatures.

2.2.3.3 Generation of Charge Carriers at the Anode and Cathode (Electron Emission Mechanisms)

Metallic contacts are full of free electrons, which can move freely within the metallic body. Under normal conditions, they cannot leave the contacts, because there is an energy barrier between metal and vacuum, which is called the *work function of the metal*. If some energy is transferred to the electrons, e.g. by heating up the contact (*thermionic emission*), or the barrier is reduced, e.g. by application of an electrical field (*field emission*), or a combination of both (*thermo-field emission*), the electrons can leave the contact surfaces and contribute to the flow of current through the arc from one electrode to the other.

Richardson-Schottky equation formulates the thermionic emission of electrons from a metallic surface [4]:

$$J_{TE} = AT^2 \exp\left(-\frac{e\varphi - \alpha E^{1/2}}{kT}\right) \tag{2.6}$$

where $A = 120.4$ [A/cm^{2}K^2], $\alpha = 3.8 \times 10^{-4}$. φ, k, E and T are the work function of the contact material [eV], Boltzmann constant, the electric field at the contact surface [V/cm] and the cathode temperature [K], respectively. J_{TE} [A/cm^2] is the thermionic emitted current density.

In case of field emission, the electrons tunnel through the energy barrier (*tunnelling effect*). If the barrier is reduced by application of an electric field, the probability of presence of electrons outside the metallic body increases and so does the emitted current density. The field emission current density is quantitatively described by Fowler-Nordheim equation [5]:

$$J_{FE} = 1.41 \times 10^{-6} \frac{E^2}{\varphi} \exp\left(-\frac{6.85 \times 10^7 \varphi^{\frac{3}{2}}}{E} \theta(y)\right) \tag{2.7}$$

where J_{FE} [A/cm^2] is the field emission current density and E [V/cm] the electric field on the contact surface, respectively. $\theta(y)$ is a function of $y = 3.62 \times 10^{-4} E^{1/2} \varphi^{-1}$ and it is approximately equal to $0.95 - 1.03$ times of y^2 [6].

In many cases, the cathode surface is hot and at the same time, an electric field is applied. The current densities are much higher than both thermionic and field emission currents. A method proposed by Murphy and Good [7] can be used to calculate the thermo-field emission current densities.

Electrons are mainly supplied to the arc column by the cathode. If the cathode is made of a metal with a high boiling point, e.g. tungsten, molybdenum or zirconium, the cathode surface may reach high enough temperatures for thermal emissions to occur. Thermal emission becomes important for temperatures greater than 3500 K, and a typical current density is 100 A/mm². Other mechanisms work for cathode materials with boiling points lower than the arc's temperature. In such cases, it is believed that field emission or a combination of thermal emission and field emission are the most important mechanisms.

The anode may be either active or passive. A passive anode does not supply electrons to the arc. As the arc temperature decreases towards the anode, the number of thermal ionisations in the arc decrease and the number of recombinations increase. This leads to a decrease of the electron density. To maintain the electric current, the electrons are accelerated to greater velocities. Consequently, there is an increased possibility for impact ionisations. In order to accelerate the electrons, a stronger electric field is needed. This is created by the net negative charge density in the anode region and the associated anode voltage drop V_a, see Fig. 2.9.

In an active anode, the surface temperature is so high that material evaporates. The metal vapour may then be ionised resulting in an arc burning in a mixture of gas and metal vapour.

The electron emission mechanisms are the main charge carrier generation mechanisms in case of low pressure (vacuum) switching devices as explained in Sect. 2.4.

2.2.3.4 Formation of Negative Ions

Negative ions are created when neutral particles capture and combine with free electrons. Such a process takes place for some molecules and atoms where the total potential energy of the electron and neutral particle becomes lower when they are combined. Consequently, energy is required to separate them. This energy is called the *electron affinity* and is generally measured in electron volts (eV). The electron affinity varies from zero for the inert gases to 3.9 eV for fluorine. It is greatest for elements lacking one electron in their outermost shell. These elements form electronegative gases, as is the case with sulphur and fluorine in SF_6.

The probability of forming a negative ion decreases with increasing electron velocity, i.e. with increasing temperature and electric field. The average time it takes before a free electron becomes attached is referred to as the *attachment time*. At room temperature, it is infinite for inert gases, about 6×10^{-7} s for air and about 4×10^{-9} s for SF_6.

2.2.3.5 Recombination

When a gas or plasma contains a mixture of negatively and positively charged particles, these can recombine to form neutral particles. They then have to reduce their combined energy, either as kinetic energy (collisions with solid surfaces, three-particle-impacts) or by radiation. Exchanging energy with the surroundings takes some time, and the recombination probability is therefore greater when particle velocities are low (i.e. low temperatures).

Among the many possible recombination processes are:

- Recombination at solid surfaces (of less importance in high pressure arcs, important in vacuum arcs).
- Recombination of positive and negative ions (high probability as both types of particles have relatively low velocities).
- Recombination of electrons and positive ions (relatively low probability as the electrons move much faster than the ions).

The recombination rates are very sensitive to changes in pressure and temperature.

2.2.4 Charge Carrier Dynamics

In the previous subsection, the relevant generation and loss mechanisms of the charge carriers were discussed. The generated charge carriers are subjected to different motion mechanisms due to the external applied forces (e.g. by an electrical field) or because of their internal energy. The directional motion of charged particles (*drift*) results in a net current flow through the switching arc, while their random motion cause them to spread or redistribute (*diffusion*).

2.2.4.1 Motion of Charged Particles in an Electric Field (Electrical Conductivity)

As previously explained, charge carriers for carrying the current in an electric arc are mainly generated in three different ways:

- Electrons are emitted from the cathode.
- Electrons and positive ions are formed by ionisation of gas molecules or atoms, i.e. one or more electrons are detached and the gas molecule or atom becomes a positive ion.
- Negative ions are formed in electronegative gases through electron capture.

When such a mixture of positive and negative charge carriers exists in an electric field, the carriers will in average move parallel to or in opposite direction of the field, depending on their polarity. The mean velocity is called the drift velocity u and can be expressed as:

$$u = \mu E \tag{2.8}$$

where μ is called the mobility of the charge carrier and E is the electric field. This movement of charge leads to an electric current density J given as:

$$J = eE\left(\mu_e n_e + \sum_{k=1}^{N} k\mu_k n_i^k\right) = \sigma E \tag{2.9}$$

The electrical conductivity of the gas can be expressed based on the densities of charge carriers and their mobility as:

$$\sigma = e\left(\mu_e n_e + \sum_{k=1}^{N} k\mu_k n_i^k\right) \tag{2.10}$$

where σ is the electrical conductivity, μ_e and n_e are mobility and density of electrons, μ_k and n_i^k are mobility and density of the k-fold ionised ions.

The arc column is generally approximately charge neutral, i.c. $n_e \approx \sum_{k=1}^{N} k n_i^k$. The mobility of ions is very much smaller than that of electrons (typically a factor of 10^3), because their mass is much larger than the electron mass. This implies that the electron density of the ionised gas mainly determines its electrical conductivity:

$$J \approx eEn_e\mu_e = \sigma E \tag{2.11}$$

where the electric conductivity σ is:

$$\sigma = e \cdot n_e \cdot \mu_e \tag{2.12}$$

e is the elementary charge. In strongly electronegative gases, the electrons quickly form negative ions with approximately the same low mobility as the positive ions. The conductivity then becomes:

$$\sigma = 2e \cdot n_i \cdot \mu_i \tag{2.13}$$

2.2.4.2   Diffusion and Ambipolar Diffusion

Concentration gradient of a certain particle type in the arc causes a flow of particles from regions of high concentration to regions of low concentration; this phenomenon is called *diffusion* (see Fig. 2.7).

Fig. 2.7 Schematically
shown flux of particles from
high density to low density
regions

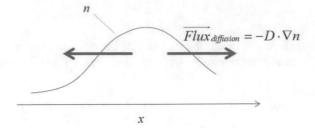

In systems containing many particle types, each type diffuses with a flux proportional to the concentration gradient of that particle type:

$$i_a = -D_a \text{ grad } N_a \qquad (2.14)$$

where i_a is the flux per volume and time of particles of type a, D_a is the diffusion coefficient for particle type a in the relevant medium, and N_a is the particle concentration.

When considering electron and ion transport in a cylindrical arc, the contributions from the electrons to the electric current density in radial direction can be expressed as

$$j_e = eD_e \text{grad } N_e + eN_e\mu_e E \qquad (2.15)$$

and in a similar way for the positive ions

$$j_i = -eD_i \text{grad } N_i + eN_i\mu_i E. \qquad (2.16)$$

The total current density in radial direction must be equal to zero under stationary conditions

$$j_e + j_i = 0. \qquad (2.17)$$

Furthermore, $N_e \approx N_i = N$ *(for singly ionized atoms)*. By inserting this into (2.14) and (2.15), E is eliminated and the following equation is found:

$$j_i = -j_e = -eD_{am} \text{ grad } N \qquad (2.18)$$

where D_{am} is called the ambipolar diffusion coefficient, given as:

$$D_{am} = \frac{D_i\mu_e + D_e\mu_i}{\mu_e + \mu_i}. \qquad (2.19)$$

The Einstein relation gives the relationship between the mobility and the diffusion coefficient:

$$\frac{D}{\mu} = \frac{kT}{e}.$$ (2.20)

In a high pressure arc an equal particle temperature is generally assumed, thus $T_e = T_i$. The electron mobility is usually far greater than the ion mobility because ions are much heavier, i.e., $\mu_e \gg \mu_i$. By inserting this into the (2.19), the ambipolar diffusion coefficient becomes:

$$D_{am} \approx 2D_i.$$ (2.21)

This expression shows that the diffusion coefficient of the heavy and slow ions determines the speed at which the radial transport of heat and charge carriers from the arc core occurs. The movement of the normally faster electrons is in this case greatly influenced by a radial electric field in such a way that the particle flux (and the associated radial heat and charge transport) is mainly determined by the drift velocity of the ions.

Depending on the medium, in which the arc is formed, only some of the mechanisms explained in Sects. 2.2 and 2.4 are dominant. In power switching devices, the current carrying contacts are normally surrounded either by insulating liquids like oil or by insulating gases like air or SF_6 or even by vacuum.

By initiation of a switching arc in insulating liquids, the liquid quickly evaporates and the arc burns in a gas bubble, so that the behaviour of the switching arc in devices with insulating liquids is very much like as in those with insulating gases. Thus, as far as the arc behaviour is concerned, all switching arcs in power switchgear can be divided into the following two categories:

- **High pressure arcs**: this includes all switching devices with insulating liquid or gas (oil breakers, gas breakers).
- **Low pressure arcs**: this is related to vacuum switching devices, where the switching arc is a metal vapour arc.

The properties and characteristics of these two arc types will be discussed in more detail in the following sections.

2.3   High Pressure Switching Arc

When considering high pressure switching arcs, it is common to distinguish between *static* and *dynamic* arcs. A static arc can be established by using a DC circuit. A DC source is connected in series with a resistance and a pair of contacts. At first, the contacts are in closed position and current passes through the circuit. By separating the contact members, an arc ignites and burns across the contact gap. When all transient phenomena have faded out and stationary conditions have been reached, a static arc is obtained.

In cases, where the current or the cooling of the arc varies with time, the arc is said to be dynamic. The voltage drop in a dynamic arc at a given time does not only depend on the instantaneous value of the current, but also of the history, e.g., the magnitude of the current and environmental conditions as they were just prior to the voltage measurement. As it will be seen later, the reason is that the electrical conductivity of the arc is highly dependent on the arc temperature and cross-section. If the temperature is to be changed instantaneously, a certain mass has to be heated or cooled, and this cannot be achieved unless infinite amounts of power are transferred. In practice, the cooling is approximately constant over a short period of time, whereas the current of the circuit determines the input power. Consequently, the thermal inertia causes the arc to "remember" for a short time the amplitude of the current that just has passed.

AC switchgears are always dealing with dynamic arcs. However, the physical properties and conditions of the dynamic arc are very complex, and in simplified descriptions, it is generally assumed that the arc, within a time interval, behaves as a static one. Static arcs are, therefore treated first.

2.3.1 Static Arc Characteristic

The relation between the current in the arc and the voltage drop between the contacts in a static arc generally has a characteristic shape. This is called the *static arc characteristic* and is shown in Fig. 2.8.

The relation shown in Fig. 2.8 applies to a freely burning arc, i.e. an arc burning in a gaseous medium where there is no interaction or influence with the walls or other parts of the interruption chamber. Vacuum arcs show a very different current–voltage relationship.

The static arc characteristic is highly nonlinear. At low currents, i.e. tens of amperes, the voltage drop across the arc decreases with increasing current. For

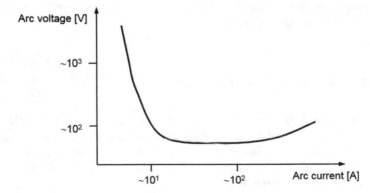

Fig. 2.8 The static arc characteristic for an arc burning in a gas gap at atmospheric pressure or greater (schematic). The scaling of the axes is approximate and varies greatly with gas type and electrode material

somewhat higher currents, the voltage across the arc is fairly constant and independent of the current amplitude. At high currents, the voltage increases somewhat
with increasing current.

The arc voltage varies quite a lot from one gas to another and is also dependent
on the electrode material and the arc length. Typical values for the flat part of the
characteristic are from a few hundred to a few thousand volts.

Figure 2.9 shows a schematic drawing of the longitudinal cross-section and
electrical potential distribution of an arc.

The electric arc is divided into three main regions:

- The cathode region with a cathode voltage drop V_c. A typical value of V_c is
  20 V.
- The arc column with an approximately constant electric field, typically
  10 V/cm, and thus with a relatively low contribution to the total voltage drop if
  the arc is short.
- The anode region with an anode voltage drop V_a, typically 3 V.

The cathode side has more distinct arc foot points than the anode. In short arcs
most of the voltage drop occurs close to the electrodes, as shown in Fig. 2.9; the
increase in potential is substantially smaller along the arc column.

The electric current in a burning arc is carried partly by electrons and partly by
positive ions, with the direction of movements as shown in Fig. 2.9. Most of the
current is carried by electrons, which in part are generated by ionisation of gas
molecules in the cathode area and in part emitted from the cathode surface. The
space charge distribution in the electric arc can be found by considering the distribution of the electric potential along the arc. The relationship between electric
flux density D and the space charge density ρ is:

$$\mathrm{div}\boldsymbol{D} = \rho. \tag{2.22}$$

Fig. 2.9 A cross-section of a
stationary arc (*top*) and the
corresponding potential
distribution (*bottom*)

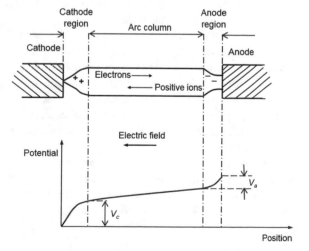

Electric flux density can be expressed by the electric field E and the electric potential ϕ by the following equations:

$$D = \varepsilon_r\varepsilon_0 E \quad E = -grad\ \phi. \tag{2.23}$$

Combining (2.22) and (2.23), and assuming constant permittivity, $\varepsilon_r\varepsilon_0$, gives:

$$\frac{d^2\phi}{dx^2} = -\frac{\rho}{\varepsilon_r\varepsilon_0} \tag{2.24}$$

This implies that there is a net negative space charge in areas, where the second derivative of $\phi(x)$ is positive (i.e. where $\phi(x)$ curves upwards). Figure 2.9 shows that this occurs in the anode region. Similarly, it is evidently a net positive space charge close to the cathode.

The second derivative in the arc column is zero ($\phi(x)$ is approximately linear), which indicates that there is no net space charge in this region. In other words, the density of electrons and (singly charged positive) ions is equal. Thus, the plasma contains many free charge carriers, but when observed as a whole from the outside it appears uncharged or neutral.

2.3.1.1 Particle Density

In switching devices with high pressure arcs, thermal ionisation is the most important charge generation process. Impact ionisation can be of relatively greater importance near current zero crossing. Ionisation due to radiation plays normally only a minor role. In a gas or an arc at stationary conditions, there is an equilibrium between the dissociation, ionisation and recombination processes. These processes determine the concentrations of the different particles. By only considering dissociation and thermal ionisation processes, various particle densities can be calculated using (2.2) and (2.4). The outcome of such a calculation is presented in Fig. 2.10 for nitrogen at constant pressure.

SF_6 decomposes in several steps (SF_4, SF_2 etc.). The processes and conditions then become very complex, and the calculated particle densities are not very accurate, particularly at higher temperatures.

The charge carrier concentration in nitrogen varies strongly with temperature as shown in Fig. 2.10. Similar conditions are found in other gases being used in switching equipment. Hence, the electrical conductivity of an arc varies over a huge span. Figure 2.11 shows the electric conductivity of air as a function of temperature.

As shown in Fig. 2.11, the arc is an excellent electric insulator (comparable to glass and porcelain) for temperatures less than a few thousand kelvin. By increasing the absolute temperature one decade, from 500 to 5000 K, the conductivity increases with as much as 14 orders of magnitude [9], approaching the conductivity

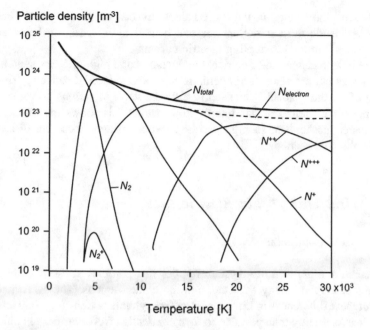

Fig. 2.10 Particle densities for nitrogen at different temperatures and atmospheric pressure [8]

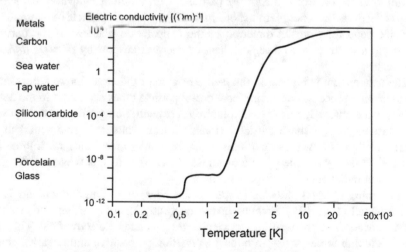

Fig. 2.11 Electrical conductivity of air at atmospheric pressure as a function of temperature. Note that both axes have logarithmic scales. Examples of materials at the different conductivities are given along the vertical axis

of metals. No other substance is capable of changing its electric conductivity this much. The conductivity span of semi-conductors in thyristors and diodes is considerably less, at most 8–9 orders of magnitude.

Moreover, the change in the arc conductivity can occur very fast. These two properties are the most important reasons behind the electric arc being such an excellent medium for interrupting electric current.

To be able to change the electrical conductance of a high-pressure switching arc abruptly near current zero, its temperature has to be reduced drastically. Therefore, *cooling* of the arc channel is realised in different ways in various power switching components with liquid or gas interruption media. This shows the importance of heat transfer mechanisms in high pressure switching arcs; these are discussed in detail in the next section.

2.3.2 Heat Transport in Electric Arcs

2.3.2.1 Heat Conductance

A temperature difference in a system implies that particles in the warmer regions have higher kinetic energy than those in the colder regions. The system seeks to level out these differences in kinetic energy through interactions (e.g. impacts). This is how heat or energy transport due to *heat conduction* from warmer to colder areas in an arc occurs.

Moreover, in case of diffusion of particles with different temperatures, such a process leads to an energy transport in the direction of particle movement, or opposite, depending on the direction of the temperature gradient. If the particles flow from a warm to a cold region, heat or energy transport by *particle diffusion* thus occurs.

The temperature is highest in the centre or core of the arc column. Dissociation and ionization processes occurring here create particle types not found in the colder regions at the outer edge of the arc column. A concentration gradient is thus present and causes a flux of these particles (ions, electrons, molecule fragments) in the radial direction. When these particles reach the colder areas, they recombine and emit their dissociation energy and/or ionisation energy. In this way particle diffusion causes radial energy transport in the arc.

Concerning charged particles (electrons and ions) certain restrictions apply for such a particle flux. If differences in the concentration of positive and negative charge carriers exist, the resulting space charges generate an electric field. This field yields Coulomb forces acting in opposite direction on positive and negative charge carriers. The direction of these forces is such that they seek to prevent space charge formation, i.e. to maintain charge neutrality.

Based on the descriptions of the different heat conduction mechanisms in the arc plasma, it is obvious that the thermal conductivity varies strongly with temperature. Without consideration of dissociation, ionisation and chemical reactions, the thermal conductivity coefficient of a gas increases slowly with temperature, but as discussed in Sect. 2.2.2, in a high pressure arc many reactions like dissociation and ionisation take place. This results in a very non-linear temperature dependent behaviour of the

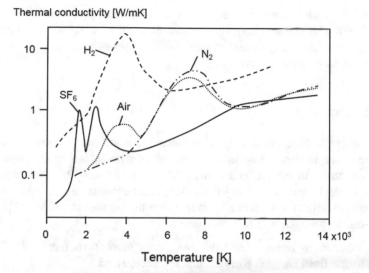

Fig. 2.12 Thermal conductivity as a function of temperature for gases at elevated temperatures [10]

thermal conductivity. This relationship is shown in Fig. 2.12 for a few relevant gases. In case of gas mixtures, e.g. air, or complex molecules with different dissociation stages, a number of peaks at different temperatures can be identified.

The extremely high thermal conductivity of molecular gases near their dissociation temperatures can be simply explained by considering the schematics depicted in Fig. 2.13. The diffusion of molecules to the higher temperature region results in increased dissociation, i.e. consumption of dissociation energy W_D, and in an accompanying large heat flux (q_D) [11]:

$$q_D = -W_D \cdot D_m \nabla n_m \qquad (2.25)$$

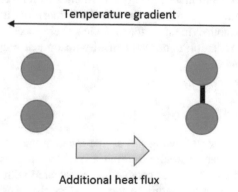

Fig. 2.13 Contribution of dissociation process to thermal conductivity

Here, n_m and D_m are the molecules density and the diffusion coefficient, respectively. This can be interpreted as an enhancement of thermal conductivity at those regions. Hence, higher dissociation energies may enhance the heat conduction in high pressure switching arcs.

2.3.2.2 Radiation

Electromagnetic radiation is of great importance for the energy transport within an arc. Moreover, radiation may also provide clues with regard to the physical processes occurring in the arc, and a great deal of the existing knowledge about the physics of the electric arc comes from studying the radiation.

Plasma in an arc can emit or absorb radiation through several different processes. These can be grouped into two categories:

- Radiation from atoms going from one energy level to another.
- Radiation from charged particles that are accelerated.

When analysing the light emitted from an arc, it is generally found that it consists of a *line spectrum* superimposed on a *continuous spectrum*.

The line spectrum is due to transitions between fixed energy levels in the atoms, ions or molecules, and is directly related to excitation processes. With the particle densities normally seen in high pressure switching arcs (not vacuum arcs), practically no radiation is emitted from the arc for transitions from an excited level to the ground state. In such transitions, the absorbing atoms will enter into the same energy state, as was the original emitting atom. The light quantum therefore diffuses around in the plasma and has a low probability of escaping. This is called *optical trapping*. This radiation may be of considerable importance for the energy transport internally in the arc if the photons diffuse from regions of high radiation intensity to regions of low radiation intensity.

The continuous spectrum originates in several processes, which can be divided into "free-bound" and "free-free" transitions. "Free-bound" transitions are transitions where a free electron of a certain kinetic energy recombines with an ion and the combined particle enters a discrete energy state. In "free-free" transitions an electron leaves one free state and enters another free state, for example by being accelerated by a nearby ion.

The continuous radiation is not associated with optical trapping and can rather easily escape the plasma.

2.3.2.3 Convection

Heat can be taken up by a flowing gas at one location and released at some other colder location (by heat conduction). Such a heat transport is referred to as *convection* and is, in contrast to heat conduction, associated with a transport of mass. The gas flow may origin in heating of the gas, since the density of a warm gas is

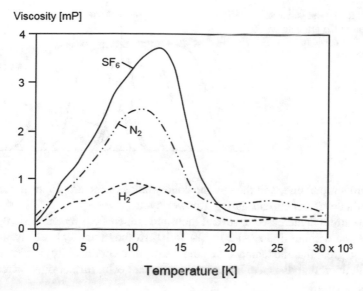

Fig. 2.14   Viscosity as a function of temperature for some gases [12]

lower than for a cold gas, and warm gas will rise and be replaced by gas flowing from colder areas. This is called *self convection*. However, in switching equipment the convection is generally due to a blowing of cold gas from an external location onto the arc. This is called forced convection or forced cooling, and is in many switchgear designs essential in extinguishing the arc at current zero.

Hence, convective cooling is closely related to flow of gases. The flow is laminar at low gas velocities and turbulent at higher velocities. The turbulence has two important aspects. It causes an increased flow resistance, but also a more efficient cooling. Whether a flow is laminar or turbulent depends, among other things, on the viscosity of the gas. The viscosity is very temperature dependent, as shown in Fig. 2.14.

Due to the temperature dependency of viscosity, the flow inside an arc can be laminar whereas the flow around the arc can be turbulent. This plays a significant role in air blast and SF_6 circuit breakers where the arc is cooled by a gas flow.

2.3.3   Temperature Distribution in an Electric Arc Column

The detailed knowledge concerning temperature distribution inside the arc of a switching device is still far from complete, but a good semi-quantitative understanding exists. The temperature profile through the arc cross-section is approximately bell-shaped at low currents (the declining part of the static arc characteristic) with a maximum in the centre. As long as the currents are relatively low, the

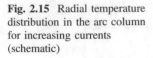

Fig. 2.15 Radial temperature distribution in the arc column for increasing currents (schematic)

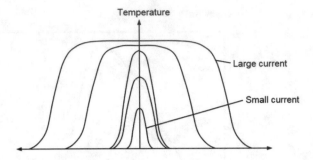

maximum temperature, and thereby the conductance of the arc, follows changes in the amplitude of the current during a power cycle.

If the current is increased to several thousand amperes or greater, the temperature does not rise without limits. At $20 - 30 \times 10^3$ K, the radiation losses from the arc core are so great that the temperature (and the electric conductivity) varies relatively little with the radial distance. Increasing the current even further only increases the arc cross-section.

Hence, in general terms, at low currents the arc temperature changes in line with changes in the current, whereas at large currents the maximum temperature is fairly constant across the cross-section. This is shown schematically in Fig. 2.15.

The arc diameter may reach a few centimetres for very large currents. The arc cross-section plays a role when designing nozzles for large circuit breakers. Here the arc is drawn through a nozzle in presence of an intense gas blast.

2.3.4 Dynamic Arcs

When the current in the arc changes rapidly, the arc voltage no longer follows the static characteristic. This is because the arc temperature and cross-section are not able to adjust instantaneously to the values corresponding to the new current value. The arc has a certain thermal inertia since a change in current leads to heating or cooling of matter. If, for example, the current follows a step function, the arc voltage will at first take a higher value, and then gradually decrease to the value corresponding to the static arc characteristic. This is shown schematically in Fig. 2.16.

The arc resistance is mainly determined by the temperature, which cannot be abruptly changed. Initially, the voltage drop is therefore directly proportional to the step in the current. The voltage waveform can be approximated by an exponential decay. The arc *time constant* can be defined as the time constant of the exponential change in arc voltage following from a step in current. Measured values for such a time constant for some gases are listed in Table 2.2.

Studies show that the arc time constant is far from a constant. It not only depends on the test conditions and the method of measurement, but also on the magnitude of the current [13].

Fig. 2.16 Voltage drop
(*solid*) of an arc exposed to a
current that follows a step
function (*dashed*)

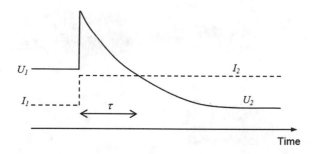

Table 2.2 Time constants
measured in a 1 A arc burning
in a 19 mm tube [13]

Gas	Time constant (μs)
SF_6	0.8
O_2	1.5
CO_2	15
Air	80
N_2	210
H_2	1

   The thermal inertia causes the current–voltage relationship to depend on how
fast the current is changing. Figure 2.17 shows arc characteristics for the static case
and for two dynamic cases.

   As Fig. 2.17 shows, the current–voltage relationship has a course similar to a
hysteresis loop where the loop narrows as the frequency increases.

   For interrupting currents, it is advantageous to have a small arc time constant.
This makes it easier to quickly change the arc temperature and thereby its electric
conductance. However, the time constant is not an unambiguous measure of
whether or not a gas is well suited as an arc quenching medium. SF_6 is, for example,

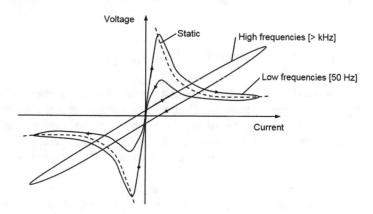

Fig. 2.17 The static arc characteristic, the arc characteristics for low frequencies (typically 50 Hz)
and for high frequencies (several kilohertz)

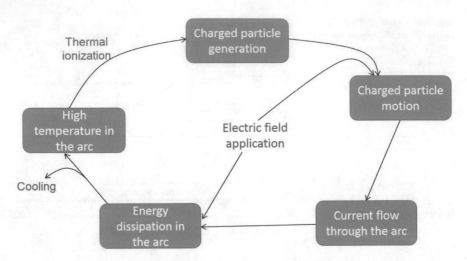

Fig. 2.18 Interrelation between important mechanisms at work in high pressure switching arcs

a far better interrupting medium than H_2 although their time constants are approximately equal, as shown in Table 2.2.

Important mechanisms in high-pressure switching arcs and their interrelations are summarized in Fig. 2.18. Charged particles responsible for current flow after opening the contacts are mainly produced by thermal ionization of the gaseous medium (see Sect. 2.2.3). The charged particles are accelerated by the electric field within the arc, and as the result of their motion a current is generated. The energy dissipation in the arc is proportional to the current flowing through the arc. The dissipated energy partly contributes to the temperature increase of the plasma. If the cooling mechanisms are so efficient that much larger amounts of power are drawn out of the arc than generated by the current flow, the temperature and consequently the conductivity of the medium decrease, and current can be successfully interrupted.

2.4   Low Pressure (Vacuum) Switching Arc

In certain switching devices, the switching arc is formed in an environment of very low pressure. In those cases, there are almost no gas atoms present and therefore, ionization of the very few gas atoms does not provide enough charge carriers to carry the electric current across the contact gap. In these devices, electron emission mechanisms at the metal contact surfaces are supplying the necessary charge carrier.

The electron current emitted from the contact surface may locally produce very high temperatures, which in turn contribute to further increase of current density. If the temperature of the contact surface in small areas exceeds its melting temperature, these areas eventually explode and a mixture of electrons, ions and metallic neutral atoms are injected into the contact gap. This mechanism is known as *explosive electron emission*. By generation of a plasma near the contact surface, the work function is reduced to zero, so that the electrons can easily flow outwards from the contact. This is mainly valid for formation of emission centres in an open contact arrangement. During contact separation of vacuum switching devices, as explained in Sect. 2.2.1, excessive thermal losses lead to formation of a first emission centre. As the electrodes are moving further apart, the initial, single arc foot point quickly splits into several foot points. This only occurs at the cathode and these micrometre-sized foot points are called *cathode spots*.

As shown schematically in Fig. 2.19, the metal atoms are explosively accelerated from the cathode spots into the switching gap. They are ionised during the first few 100 nm by the emitted electrons. Each cathode spot has a cone-shaped structure. Near the cathode surface a positive space charge region (*cathode sheath*) is formed. In a distance of a few micrometres from the cathode surface, a metal vapour plasma is formed; its density distribution is very non-homogenous and reduces strongly in the direction towards the anode (*inter electrode plasma*). The number of emission centres is very much dependent on the current amplitude and the contact material. The average current per cathode spot is in the range of some tens to 100 A depending on the cathode material (see Table 2.3). In case of low current arcs, the anode is passive and acts only as a sink of electrons and there are many cathode spots on the cathode surface, which are moving very fast all over the contact surface. The cathode spots are the only powerful light source in the gap between the contacts. There is no confined or well-defined arc column, and there is no visible foot point of the arc on the anode side. A characteristic weak, diffuse light is emitted from the arc between the electrodes, and this type of arc is thus usually called *diffuse mode vacuum arc*.

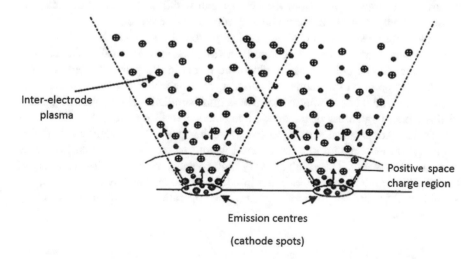

Fig. 2.19 Schematics of emission centres structure on the cathode surface

Table 2.3 Cathode spot erosion rate, average cathode spot current, arc voltage in diffused mode and chopping current level for different cathode materials [14]

Cathode material	Cathode spot erosion rate (μg/C)	Average current per cathode spot (A)	Arc voltage (V)	Average and maximum chopping currents (A)
Cu	140, 150	75–100	20	15, 25
Cr	115, 130	30–50	18	7, 16
Ag	22–27, 40	60–100	18	3.5, 6.5
W	55, 62, 64	250–300		16, 350

This arc mode is associated with low contact erosion (see Table 2.3), and the switching gap recovers very fast after current zero, making an interruption of the current easy.

As the radiation from the cathode spots is the dominating loss mechanism and an increase in current simply leads to formation of more cathode spots, the arc voltage in a diffuse mode vacuum arc is almost independent of both the distance between the contacts and of the current magnitude, see Fig. 2.21. Typical values of current per cathode spot and arc voltage as well as cathode spot erosion rates for different materials are listed in Table 2.3. The arc voltage is very low in comparison to that of high pressure switching arcs. Consequently, the energy dissipation during an interruption becomes relatively small, which is a clear advantage.

In case of vacuum (metal vapour) arcs, the recovery process of the switching gap takes place mainly by diffusion of the generated charge carriers out of the switching gap. As explained in Sect. 2.2.4.2, the diffusion dominated particle flux is very dependent on the gradient of its concentration. The gradient of particle concentrations is extremely high in case of metal vapour arcs, and consequently the flux of the particles out of the switching gap becomes large. It results in an almost complete removal of charge carriers from the contact gap within a few microseconds after cease of supply of new charge carriers at current zero crossing.

If the current exceeds a threshold, typically around 10–15 kA depending on the contact material and geometry, the arc behaviour changes drastically. The cathode spots gather in one point, a well-defined arc forms in the gap and there is also a distinct foot point on the anode, an *anode spot*. This is called a *constricted mode vacuum arc*, and it has an appearance that is quite similar to a high pressure gas arc. In this mode, large areas of the anode and cathode surface melt and both contacts can provide some metal vapour to the switching gap. The molten and very hot contact surfaces continue to release metal vapour to the switching gap even after current zero crossing. This results in very high contact erosion and slow recovery of the switching gap. Therefore, entering into this arc mode is clearly undesirable, and should preferably be avoided during the operation of vacuum switching components.

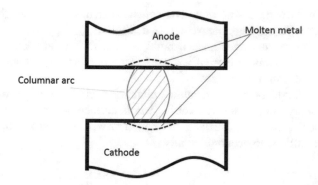

Fig. 2.20 Schematically shown structure of a constricted vacuum arc

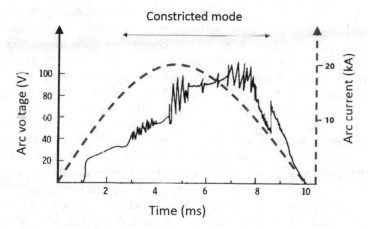

Fig. 2.21 Arc voltage of a vacuum interrupter at different currents

Figure 2.20 shows the constricted vacuum arc schematically. Dissipations contributing to the melting of large area on the anode and cathode surfaces are partly due to the radiation losses of the constricted mode arc, but mainly due to the losses occurred by current flow through the arc near electrode regions. A significant voltage drop exists near cathode and anode. This implies that energies of $\int i(t) \cdot V_c dt$ and $\int i(t) \cdot V_a dt$ are dissipated close to cathode and anode surfaces.

Due to establishment of new arc foot points, the arc voltage changes sporadically, but is in average much higher than that of a vacuum arc in diffuse mode, as illustrated in Fig. 2.21.

When the current decreases towards the end of the half cycle, the arc returns to its more harmless diffuse mode. The cathode spots die one by one, and finally, there is only one spot left. This abruptly ceases to exist just before the natural current zero crossing, and the current is cut off. This is called *current chopping* and is a typical vacuum interrupter phenomenon, although it can occur under certain conditions in other types of switchgears as well. The contact material determines the chopping current level in a vacuum interrupter (see Table 2.3). Current chopping is an unwanted phenomenon since such a rapid change in current may lead to over-voltages. A low chopping current is therefore favourable when choosing contact materials for vacuum interrupters.

Thus, the control of arc in vacuum switching devices means to distribute the dissipated energy homogenously over the whole surface of the contacts to prevent large area melting of contact surface. In Chap. 4, it will be described how magnetic fields can be utilized to accomplish this.

2.5 Modeling of Switching Arcs

The switching arc is the key element, which enables interruption of currents in mechanical switching devices. Therefore, methods contributing to a better under-standing its behaviour have received considerable attention. Depending on the perspective and goals of the research, numerous models with different complexity levels and for various applications have been developed for switching arcs.

In some cases, e.g. if the switching transients are to be modelled and the rated voltage of the network is much higher than the arc voltage of the switching com-ponent, the switching arc can be modelled as an ideal switch opening at current zero, i.e. zero resistance in closed and arcing state, and zero conductance in open position.

In case the interaction between the switching arc and components of the sur-rounding network is of importance, the simplest models describe the arc as a two-port element with a nonlinear interrelation between arc voltage and arc current. H. Ayrton suggested one of the most famous arc models of this kind [15], when she investigated the characteristics of the arcs burning between carbon electrodes. She postulated:

$$u_{arc} = a + b \cdot l + \frac{c + d \cdot l}{i_{arc}} \tag{2.26}$$

where u_{arc} and i_{arc} are arc voltage and arc current, respectively; a, b, c and d are constants and l is the length of the arc. This equation fits to the behaviour of static arcs for currents less than a few tens of amperes (see Fig. 2.8). However, no time dependency is included, so this model cannot be used to describe dynamic arcs as those in power switchgear.

2.5.1 Detailed Physical Models

To be able to develop applicable switching arc models, physical phenomena taking place in the arc have to be considered. The most accurate but complex models are the so-called *detailed physical models*. Here, all the physical mechanisms are described by their governing equations and those equations are solved simultaneously using numerical methods to determine the time and space resolved physical parameters, such as density, velocity and temperature of different particle types present in the arc. This in turn is used to calculate the integrated, measurable quantities like arc voltage and arc current.

Depending on the interrupting media, the governing equations for the arc plasma could be very different.

2.5.1.1 High Pressure Switching Arc

For high pressure switching arcs, it is normallyassumed that a local thermodynamicequilibrium exists, and therefore all particle types are supposed to have the same temperature. In this case, a plasma is assumed being a fluid with high electrical conductivity, capable of interacting with electromagnetic fields. Hence, the governing equations of fluid dynamics, e.g. Navier-stokes equation, are solved together with the electromagnetic field equations. This formulation is known as magneto-hydrodynamic (MHD) model. Three sets of equations, namely fluid dynamic equations, electromagnetic field equations and material equations of state are considered:

- Fluid dynamic equations· These include mass, momentum and energy conservation equations.

$$\frac{\partial \rho}{\partial t} + \nabla \cdot (\rho \cdot v) = 0$$
$$\frac{\partial (\rho \cdot v)}{\partial t} = -\nabla(\rho v v) - \nabla p + \eta \nabla^2 v + j \times B \qquad (2.27)$$
$$\frac{\partial}{\partial t}(\rho H) + \nabla \cdot (\rho v H) - \nabla(\lambda \nabla T) = \frac{\partial p}{\partial t} + j \cdot E - E_{rad}$$

Here ρ and v are mass density and velocity of the fluid (i.e. the plasma). H, p and T are the total enthalpy, pressure and temperature, respectively. The total enthalpy H can be expressed as:

$$H = \frac{1}{2}v^2 + h \qquad (2.28)$$

where h is the specific enthalpy of the material and is a function of its density and temperature. The viscosity coefficient and the specific thermal conductivity are denoted η and λ. E_{rad} is the radiation losses. E, B and j are the electric field, the magnetic flux density and the current density, respectively.

- Maxwell's equations:

These are in general form expressed as:

$$\nabla \times \boldsymbol{E} = -\frac{\partial \boldsymbol{B}}{\partial t}$$
$$\nabla \times \boldsymbol{H} = \boldsymbol{j} + \frac{\partial \boldsymbol{D}}{\partial t}$$
$$\nabla \cdot \boldsymbol{E} = \frac{\rho_{elec}}{\varepsilon_0} \qquad (2.29)$$
$$\nabla \cdot \boldsymbol{B} = 0$$

In a switching arc (plasma), as mentioned earlier, there is almost no net charge and therefore, $\rho_{elec} \approx 0$. In addition, the impact of the time derivative of the magnetic flux density $\left(\frac{\partial \boldsymbol{B}}{\partial t}\right)$ is in many cases negligible, because the frequency range of interest is not so high. This implies that the electrical field is rotational free and therefore, an electrical potential φ may be defined within plasma arc. The electric potential is calculated by solving the Laplace's equation. The current density can then be expressed as:

$$j = -\sigma \nabla \phi \qquad (2.30)$$

- Material equation of state:

(2.27) contains thermodynamic quantities of the interrupting medium. These are in general not constant, but functions of the arc density and temperature, as discussed in Sect. 2.2. The set of equations is only complete if these interrelations are taken into consideration.

The governing equations of the high pressure arc are solved together with appropriate boundary and initial conditions. Normally, numerical methods and commercial CFD (computational fluid dynamics) software are used. The output of such simulations is the spatial distribution of the gas temperature and pressure (density) at different times. Based on the gas temperature and pressure, the conductance of the switching device at different times, especially near current zero can be calculated.

2.5.1.2 Low Pressure (Vacuum) Switching Arc

In one extreme case, when considering the switching arc in low-pressure switching devices with diffuse plasma, there is no local thermodynamic equilibrium and therefore different species, like electrons, ions and neutral particles may have different temperatures. For this case, either two flow plasma models [16] or Boltzmann equation based plasma models [17] are used to describe its behaviour. In constricted mode, where an arc column exists, magneto-hydro dynamic models like those applied to model high-pressure arcs are used [18]. The output of these simulations is the current density and energy flux distribution to the contact surfaces at different times.

Contrary to high-pressure arcs, the electrodes play the dominant role in recovery process of low-pressure (vacuum) switching arcs. The energy flux to cathode and anode may result in temperature increase of the contact surfaces and consequently to metal vapour supply to the switching gap after current interruption. This is the main reason for thermal re-ignition in vacuum interrupters. The temperature distribution at the contact surface may be modelled by considering the heat conduction through the bulk of the metallic contact.

2.5.2   Qualitative Physical Models

In order to reduce the complexity of the governing equations, one simplification would be to integrate (2.27) in one or more spatial dimensions (e.g. radial r and axial z directions). From a physical perspective, this means to divide the arc in many zones with different simplified governing equations. Those equations can then be solved using appropriate boundary conditions between adjacent zones. One of the simplest qualitative (integral) models is the so-called two-zone model [19]. In this model, the arc is split into two zones, arc column and arc surroundings, see Fig. 2.22. It is assumed that all current flows through the arc column (zone 1) and that the energy dissipation takes place in this zone. The second zone is only important for the radial transport of the generated heat out of the arc column. This method has been successfully applied to describe switching arcs in high voltage gas circuit breakers.

Fig. 2.22 Temperature profile of a high pressure arc with indication of two distinct zones

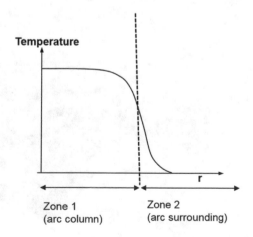

In terms of computational requirements, the qualitative methods are much more efficient than the detailed physical models and are therefore applied for arc-network interaction, in simple electrical circuits, and for simple current interruption investigations.

2.5.3 Black Box Models

In the two categories of arc models described above, the physics of the arc stands in the foreground, but in many cases, there is no need to delve into the physical behaviour of the arc. The only important fact is how the arc interacts with other components of the power network. For such cases, it is sufficient to find a relationship between arc voltage and arc current. So-called b*lack box modelling* is a way to describe voltage–current characteristics of power switching devices during their arcing time, regardless of the physical parameters of the switching arc. The scope of these three different approaches used for modelling of high pressure switching arcs is shown schematically in Fig. 2.23.

The foundation of all black box models is the fact that the energy stored in the arc Q strongly influences its conductance. Under the assumption that all cooling mechanisms of the arc contribute to reduction of the energy stored in the arc with an equivalent cooling power $P_{cooling}$, and considering the input power to the arc due to the current flow through it, the following expression can be derived:

$$Q = \int \left(u_{arc} \cdot i_{arc} - P_{cooling} \right) dt \tag{2.31}$$

The difference between different black box methods is how the dependency of the conductance of the arc (g) on the energy stored in it (Q) is expressed:

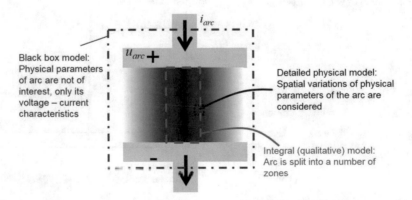

Fig. 2.23 Different approaches to modelling of switching components with high-pressure arcs

$$g = f(Q) \tag{2.32}$$

After differentiation of (2.32) and using (2.31), the following general expression is found:

$$\frac{dg}{dt} = \frac{dg}{dQ} \cdot \frac{dQ}{dt}$$

$$\frac{dg}{dt} = f'(Q) \cdot \left(u_{arc} \cdot i_{arc} - P_{cooling} \right) \tag{2.33}$$

The latter equation is the basic equation for all black box models. Note that, no assumptions are made with regard to how the arc conductance varies.

2.5.3.1   Cassie Black Box Model

If the behaviour of the arc under high currents is considered, the arc can be simply modelled as a cylindrical column with a radius dependent on the energy stored in it, see the temperature profile of the arc in Fig. 2.15. This means that the electrical conductivity of the arc column is assumed to be constant, and a change of the arc current leads only to a change in the arc column radius (i.e. the arc current density is constant as well). In this case, the relationship between arc conductance and the energy stored in arc is linear ($g = kQ$). In addition, it is assumed that the cooling power is proportional to the conductance of the arc ($P_{cooling} = U_C^2 \cdot g$). So by replacing $f'(Q) = k = \frac{g}{Q}$ and considering that $g = \frac{i_{arc}}{u_{arc}}$, the general equation of the black box models can be written as:

$$\frac{1}{g}\frac{dg}{dt} = \frac{P_{cooling}}{Q} \left(\frac{u_{arc}^2}{U_c^2} - 1 \right). \tag{2.34}$$

The ratio of the energy stored in the arc to the cooling power has the dimension of time and describes how fast the arc reacts to change of its input/output power. It is almost identical to the arc time constant introduced in Sect. 2.3. As a second simplification, this ratio is assumed constant and denoted τ_c. The resulting differential equation for what is known as the Cassie black box model [20] can then be expressed as follows:

$$\frac{1}{g}\frac{dg}{dt} = \frac{1}{\tau_c} \left(\frac{u_{arc}^2}{U_c^2} - 1 \right); \quad g = \frac{i_{arc}}{u_{arc}} \tag{2.35}$$

Thus:

$$\frac{1}{i_{arc}}\frac{di_{arc}}{dt} - \frac{1}{u_{arc}}\frac{du_{arc}}{dt} = \frac{1}{\tau_c} \left(\frac{u_{arc}^2}{U_c^2} - 1 \right) \tag{2.36}$$

In this way, a closed mathematical relationship between arc voltage and arc current is derived. The resulting model contains two parameters U_c and τ_c, both independent of time.

2.5.3.2 Mayr Black Box Model

As discussed in Sect. 2.3.3, at lower arc currents, the temperature of the arc core is very much dependent on the current amplitude. This corresponds to the temperature range where the electrical conductivity of the arc is strongly dependent on its temperature; see Fig. 2.15. As a simplification, the arc column can in this regime be modelled as a cylinder with constant diameter, and with a conductivity depending on the energy stored in the arc. In Mayr black box formulation [21], the arc conductance is assumed to be an exponential function of the stored energy in it:

$$g = g_0 \cdot \exp\left(\frac{Q}{Q_0}\right) \tag{2.37}$$

where g_0 and Q_0 are constants. The cooling power is assumed to be constant, $P_{cooling} = P_0$. Replacing $f'(Q) = \frac{g_0}{Q_0}\exp\left(\frac{Q}{Q_0}\right) = \frac{g}{Q_0}$ in (2.33), this yields the following differential equation:

$$\frac{1}{g}\frac{dg}{dt} = \frac{P_0}{Q_0}\left(\frac{u_{arc} \cdot i_{arc}}{P_0} - 1\right) \tag{2.38}$$

As in case of Cassie black box model, the ratio $\frac{Q_0}{P_0}$ has the dimension of time and represents the arc time constant. Also in Mayr formulation, this ratio is replaced with a constant τ_m. In this way, the final differential equation of Mayr black box model becomes:

$$\frac{1}{g}\frac{dg}{dt} = \frac{1}{\tau_m}\left(\frac{u_{arc} \cdot i_{arc}}{P_0} - 1\right)$$

$$or \tag{2.39}$$

$$\frac{1}{i_{arc}}\frac{di_{arc}}{dt} - \frac{1}{u_{arc}}\frac{du_{arc}}{dt} = \frac{1}{\tau_m}\left(\frac{u_{arc} \cdot i_{arc}}{P_0} - 1\right)$$

The Mayr formulation is very useful for describing the behaviour of a high-pressure switching arc near current zero, where the current amplitude is rather low and heat conduction is the dominating cooling mechanism.

The parameters of the black box models can be obtained from experimental measurements of the arc voltage for a given arc current. The parameter values are determined as those minimizing the difference between simulated and measured arc voltage.

In many cases, two parameters are not enough to give a satisfactory modelling of the arc behaviour for all current ranges. It is possible to combine Mayr and Cassie formulations, e.g. by considering the total arc conductance as the sum of partial arc conductances found by Mayr and Cassie models [22]. It is also possible to introduce new parameters, e.g. by taking into account the dependency of the arc time constant and/or cooling power on different quantities such as arc conductance and arc energy [23, 24].

2.5.3.3 Application of Black Box Model for Simulation of Arc—Network Interaction

The mathematical relationship between arc voltage and arc current, established by black box models, can be used together with circuit equations to simulate the arc voltage and arc current as functions of time in a switching device installed in a power network.

As the differential equations of black box models are nonlinear, it is in most cases not possible to find a closed form solution for the arc voltage or the arc current. Numerical methods are used to solve the set of equations. The classical Mayr and Cassie black box models are implemented in many scientific software packages like as Matlab/Simulink, and numerical methods such as finite difference method may be directly applied to solve the equations in an iterative manner.

The following example demonstrates how the differential equations of black box models can be transformed into a set of algebraic equations using the finite difference method. The resulting set of algebraic equations can then be solved numerically.

Consider the simple circuit of Fig. 2.24, where a circuit breaker is used to clear an inductive short circuit fault current. The equations of this circuit can be expressed as follows (note that the Mayr formulation of (2.39) is used to describe the voltage–current relationship of the arc):

Fig. 2.24 An exemplary circuit to study arc—network interaction using black box models

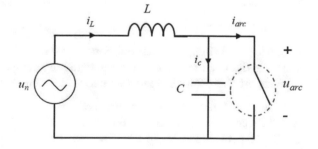

$$\frac{dg}{dt} = \frac{1}{\tau_m}\left(\frac{i_{arc}^2}{P_0} - g\right) \qquad g = \frac{i_{arc}}{u_{arc}}$$

$$i_L = i_C + i_{arc}$$

$$L\frac{di_L}{dt} = u_N - u_{arc} \tag{2.40}$$

$$i_C = C\frac{di_{arc}}{dt}$$

Replacing the time derivatives by finite difference (e.g. $\frac{dg(t)}{dt} \approx \frac{g(t+\Delta t)-g(t)}{\Delta t}$ if $\Delta t \to 0$), the above set of differential equations can be transformed to the following set of algebraic equations:

$$i_{arc}(t+\Delta t) = u_{arc}(t)g(t+\Delta t) = \frac{\left(\frac{i_{arc}(t)^2}{P_0 \cdot \tau_m} + \frac{g(t)}{\Delta t}\right) \cdot u_{arc}(t)}{\left(\frac{1}{\Delta t} + \frac{1}{\tau}\right)}$$

$$i_L(t+\Delta t) = \Delta t \cdot \frac{(u_N(t+\Delta t) - u_{arc}(t+\Delta t))}{L} + i_L(t) \tag{2.41}$$

$$i_c(t+\Delta t) = i_L(t+\Delta t) - i_{arc}(t+\Delta t)$$

$$u_{arc}(t+\Delta t) = \Delta t \cdot \frac{i_C(t+\Delta t)}{C} + u_{arc}(t)$$

Using (2.41) and starting with the initial conditions of four variables i_L, i_C, i_{arc} and u_{arc} at t_0, it is possible to calculate all circuit variable values for the times $t_0 + \Delta t$, $t_0 + 2\Delta t$, etc. In this way, all circuit variables are determined for all times steps.

In case of complex networks, the number of circuit variables and equations increase, and implementing black box models directly into circuit analysis software is a more convenient approach than manually establishing the algebraic equations.

Exercises

Problem 1—Switchgear Introduction
What are the main requirements for a switching device?

What are the main differentiators between different types of switching devices?

Problem 2—Energy Dissipation in Arc
Assume a constant arc voltage of 600 V and a sinusoidal current with the frequency of 50 Hz peak value of 50 kA. The arc burns for a half period before it is extinguished.

Calculate the total energy dissipated in the arc.

What is the maximum and mean power dissipation?

Fig. 2.25 Figure of problem 3

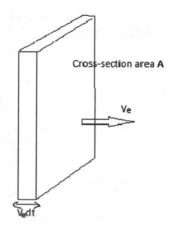

Cross-section area A

Ve

$V_e dt$

Problem 3—Derive Current Density Equation for Moving Charge Carriers
Consider the arc cross-section in Fig. 2.25.

The charge carriers in the arc with the charge q_e and the density n_e move with a velocity, v_e. Show that the current density, J, can be expressed as:

$$J = q_e n_e v_e$$

Problem 4—Conductivity in Gaseous Media
Figure 2.11 shows the relation between temperature and conductivity (in air). Why is there such a big difference in conductivity at different temperatures?

Explain in your own words the process of transition (a) from insulating to conducting state and (b) from conducting to insulating state by explaining the role of different physical processes including dissociation, ionization, recombination and heat transport.

Problem 5—Thermal Conductivity
Considering Fig. 2.26, please explain why H_2 and SF_6 have distinct peaks at in their thermal conductivity in contrast to the inert gas Argon.

Computer Exercises

In the following computer exercises, MATLAB/Simulink is used to model a switching arc and the resulting transient recovery voltage. A demonstration model that exists in the MATLAB package (power_arcmodels) will serve as the starting point for this exercise.

Problem 1
Implement the circuit of Fig. 2.27 in Simulink. Use existing Cassie and Mayr arc models to simulate the current interruption process. The parameters of the Cassie

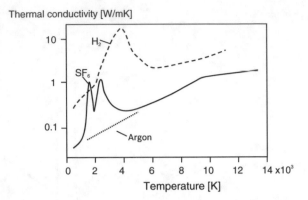

Fig. 2.26 Figure of problem 5

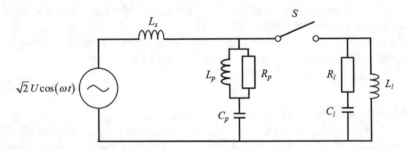

Fig. 2.27 Figure of problem 1

and Mayr models are set to, tau = 1.2 µs and Uc = 2 kV, and tau = 0.3 µs and P = 30.9 kW, respectively. The following starting values are used for different parameters:

$$L_s = 3.52\,\text{mH}, L_p = 5.28\,\text{mH}, L_l = 0.625\,\text{mH}, C_p = 1.98\,\text{mF}, C_l = 1.93\,\text{nF},$$
$$R_p = 30\,\text{W}, R_l = 450\,\text{W} \quad \text{and} \quad U = 60\,\text{kV}.$$

Explain the difference in arc voltage and current resulting from the two models. Which one is able to model a successful interruption? Which gives a more realistic arc voltage before current zero? Which one gives a realistic voltage after current zero?

Problem 2—Cooling Power and Time Constant in the Mayr Model
In this problem, only the Mayr model is considered. By double-clicking the breaker model, you can set new values to the parameters:

- Arc time constant, tau
- Cooling power, P

If the cooling power is reduced, the interruption capability of the breaker is reduced.

• Find the minimum cooling power needed to interrupt this circuit.

If the arc time constant is increased the recovery of insulation medium is slower, and the interruption capability is reduced.

• Set P back to the original, 30,900 W. Find the critical arc time constant.

Problem 3—Transient Recovery Voltage, TRV (Using the Mayr Arc Model)

(a) In this problem, a terminal short circuit fault is investigated. The short circuit is simply generated by reducing the value of the load impedance (the arc model does not allow for zero impedance between its output and ground). Replace the load RLC combination in Fig. 2.27 by a 1 mΩ resistor.

- Set cooling power, P, to 10,000 W
- Set arc time constant, *tau*, to 0.6e−6
- Run the model and find:

    – Peak of the TRV U_{peak}
    – Approximate rate of rise of recovery voltage, RRRV (given in V/μs)

(b) In a case like this, the RRRV is very dependent on the capacitance of the supply circuit (left side)

- Adjust the supply side capacitance to increase the RRRV.
- What is the maximum RRRV that this breaker can interrupt at this given current?

Problem 4—Current Limiting Circuit Breakers

The Mayr arc model gives a good description of the arc behaviour near current zero. In this problem, a series combination of Cassie and Mayr arc models as shown in Fig. 2.28 is used to model the arc voltage during the entire arcing time. The parallel resistor R_p is used to avoid numerical instabilities; its value can be set 100 kΩ, so the current flowing through this path is negligible.

Fig. 2.28 Figure of problem 4

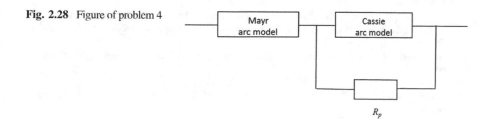

R_p

(a) Set the value of the capacitance C_p back to the original 1.98 μF.

- Run the model and find:

  - The peak current
  - The arcing time
  - The TRV amplitude and RRRV (approximate)

(b) Reduce the source voltage and the value of the source side inductor L_s by a factor 10.

- Run the model and find:

  - The peak current
  - The arcing time
  - The TRV amplitude and RRRV

(c) Change the circuit breaker separation time to 0.017 (in both models). Run the model and find:

- The peak current
- The arcing time
- The TRV amplitude and RRRV

(d) Compare the results of (a), (b) and (c) in terms of interactions between the arc voltage and the short circuit current. In which switching components are high arc voltages desirable, what are drawbacks of high arc voltages? Explain.

References

1. Chen FF (1984) Introduction to plasma physics and controlled fusion. Springer, Berlin
2. Bittencourt JA (2004) Fundamentals of plasma physics, 3rd edn. Springer, Berlin
3. Friedman A, Kennedy LA (2004) Plasma physics and engineering. Taylor & Francis Books Inc, New York, p 18
4. Richardson OW (1921) The emission of electricity from hot bodies. Longmans Green Co., London
5. Fowler RH, Nordheim LW (1928) Electron emission in intense electric fields. Proc R Soc 119:173–181
6. Bhattacharya S, Ghatak KP (2012) Fowler-Nordheim field emission. Springer, Berlin
7. Murphy EL, Good RH (1956) Thermionic emission, field emission and the transition region. Phys Rev 102(6):1464–1472
8. Lindmayer M (ed) (1987) *Schaltgeräte: Grundlagen, Aufbau, Wirkungsweise*, Springer, Berlin, p 7
9. Rieder W (1970) "Circuit breakers" In: IEEE spectrum, pp 36–43
10. Lindmayer M (ed) (1987) *Schaltgeräte: Grundlagen, Aufbau, Wirkungsweise*, Springer, Berlin, p 8
11. Friedman A, Kennedy LA (2004) Plasma physics and engineering. Taylor & Francis Books Inc, New York

12. Lindmayer M (ed) (1987) *Schaltgeräte: Grundlagen, Aufbau, Wirkungsweise*, Springer, Berlin, p 9
13. Frind G (1960) Über das Abklingen von Lichtbögen: II. Prüfung der Theorie an experimentallen Untersuchungen. Z angewandte Physik 11:515–521
14. Slade PG (2008) Vacuum interrupters: theory, design and applications. CRC Press, Boca Raton
15. Ayrton H (1902) The mechanism of the electric arc. Philos Trans R Soc Lond Ser A, pp 299–335
16. Callen JD (2006) Fundamentals of plasma physics. University of Wisconsin
17. Jenkes K, Heil B, Schnettler A (2002) Simulation of vacuum arcs in circuit breakers based on kinetic modelling. In: Proceedings of 20th international symposium on discharges and electrical insulation in vacuum, pp 634–637
18. Schade E, Shmelev L (2003) Numerical simulation of high current vacuum arcs with an external magnetic field. IEEE Trans Plasma Sci 31(5):890–901
19. Keidar M, Beilis I (2013) Plasma engineering: applications from aerospace to bio- and nanotechnology. Elsevier Inc., Amsterdam
20. Cassie AM (1939) 'Arc rupture and circuit severity: a new theory.' Report no. 102, CIGRE
21. Mayr O (1943) 'Über die Theorie des Lichtbogens und seine Löschung'. ETZ-A 64:645–652
22. Habedank U (1988) On the mathematical description of arc behavior in the vicinity of current zero. etz Archiv 10:339–343
23. Guardado J et al (2005) An improved arc model before current zero based on the combined Mayr and Cassie arc models. IEEE Trans Power Deliv 20:138–142
24. Shavemaker P, van der Sluis L (2000) An improved Mayr-Type arc model based on current-zero measurements. IEEE Trans Power Deliv 15:580–584

Chapter 3
Application of Switching Devices in Power Networks

In the previous chapters, current interruption in switchgear with mechanically opening contacts has been considered, and important parameters of the current interruption process have been introduced. It has been explained that current waveform and amplitude as well as steepness and amplitude of the transient recovery voltage constitute the critical stresses to the switching arc and have a decisive impact on whether an interruption will succeed or fail. These stresses are very dependent on the network configuration where the switching device is employed. On the other hand, switching operations may expose other components in the network to higher stresses, such as overvoltages and overcurrents.

In this chapter, different applications of power switching devices are studied. In each application, the following three matters are discussed:

- Stresses applied to the power switching device (current waveform and transient recovery voltage)
- Stresses applied to the other power components due to the switching operation (overvoltage, overcurrent, etc.)
- Methods to ensure the satisfactory performance of the switching device under different switching duties (testing methods and qualification standards).

The applications considered in this chapter include interruption of short circuit currents under various network conditions, closing operation under fault conditions, as well as energization and de-energization of different types of loads.

3.1 Interruption of Short Circuit Currents

One of the most important functions of a breaker is to protect the power system in case of a fault, which in many cases is associated with the flow of a short circuit current through a part of the network. Depending on the network configuration, the switching device may be exposed to different stresses.

© Springer International Publishing AG 2017
K. Niayesh and M. Runde, *Power Switching Components*, Power Systems,
DOI 10.1007/978-3-319-51460-4_3

3.1.1 Short Circuit Currents

Occurrence of a fault in a power network results in a change of its topology. In many cases, a fault means that a short circuit is created between one or more phases with or without connection to the earth potential. The short circuit current waveform consists of a network frequency AC and a decaying DC current component superimposed on each other.

3.1.1.1 AC Component of the Short Circuit Current

The AC part of the short circuit current can be calculated using the symmetrical component method [1]. The amplitude of AC short circuit current is dependent on the type of short circuit (whether one, two or three phases are involved) and on the neutral point treatment of the network; more precisely the impedance between neutral point and earth potential. Calculations of different AC short circuit waveforms are treated comprehensively in the textbooks on power system analysis, see for example [2], and is not considered here. Short circuit currents are found as the steady state solution of the differential equations describing the network. In the following, the most relevant concepts are briefly reviewed through some examples.

In a so-called *balanced short circuit*, the amplitude of the short circuit current in all three phases is the same, and they have a 120° phase shift relative to each other. Hence, the short-circuit currents of all three phases present a balanced three-phase system. In such cases, the three phases can be decoupled. A circuit diagram of a three-phase short circuit with earth connection in a network with solidly grounded neutral point is shown in Fig. 3.1.

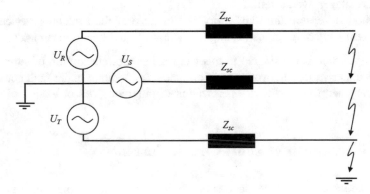

Fig. 3.1 Three-phase short circuit with earth connection in a network with solidly grounded neutral point

The short circuit current amplitude in each phase is:

$$I_{sc} = \frac{U_n}{|Z_{sc}|} \tag{3.1}$$

where I_{sc} is the amplitude of short circuit current, U_n the rated voltage of the network and Z_{sc} the short circuit impedance of each phase. In this example, the short circuit impedance contains no negative or zero sequence component.

In some cases, the fault does not affects all the three phases to the same extent, so that the fault currents flowing through different phases are not of the same amplitude. These types of faults are termed *unbalanced short circuits*. One example is a single phase earth fault, see Fig. 3.2. In this case, the other two phases are healthy. The short circuit current amplitude in the faulty phase becomes:

$$I_{sc} = \frac{U_n}{|Z_{sc} + Z_0|} \tag{3.2}$$

where Z_0 is the impedance between the neutral point and ground.

In systems with insulated neutral point ($Z_0 \to \infty$), no short circuit current is flowing through the faulted phase, during a single phase fault. In practice, due to the stray capacitances, a very small capacitive current flows through a single phase earth fault, but this can be compensated for by using an inductance (Peterson coil) between the neutral point and ground. If the neutral point of the network is solidly grounded ($Z_0 = 0$), the amplitude of a single phase earth fault becomes as large as that of a three phase short circuit current.

3.1.1.2 DC Component of the Short Circuit Current

In addition to the steady state network frequency short circuit current, the short circuit current also contains a decaying DC component. This is due to the current

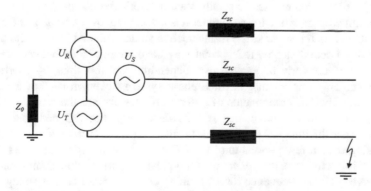

Fig. 3.2 Single phase earth fault in a network with grounded neutral point

Fig. 3.3 Single phase short
circuit in a network with
combined resistive and
inductive impedance

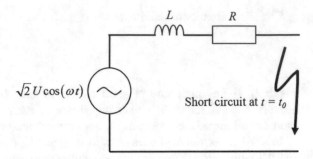

Short circuit at $t = t_0$

continuity requirement of the short circuit inductance of the network. This corresponds to the transient part of the solution of the differential equations of electrical circuits, see Fig. 3.3.

The governing differential equation for this circuit can be written as follows:

$$L\frac{di}{dt} + Ri = \sqrt{2}\,U\,\cos(\omega t) \tag{3.3}$$

Under assumption that no current flows prior to the short circuit, the following initial condition is valid:

$$i(t_0) = 0 \tag{3.4}$$

By solving (3.3) using the initial condition of (3.4), the current waveform is calculated as:

$$
\begin{aligned}
i(t) = &\frac{\sqrt{2}\,U}{R^2 + (L\omega)^2}(R\,\cos(\omega t) + L\omega\,\sin(\omega t)) \\
&- \frac{\sqrt{2}\,U}{R^2 + (L\omega)^2}(R\,\cos(\omega t_0) + L\omega\,\sin(\omega t_0)) \cdot e^{-\frac{R}{L}(t-t_0)}
\end{aligned}
\tag{3.5}
$$

The first term represents the steady state network frequency component of the short circuit current, while the second term is a decaying DC current. The amplitude of the DC part is the same as the instantaneous value of the AC part at the time the short circuit occurs. Hence, DC current may have an amplitude between zero and maximum of network frequency part depending on the time of short circuit occurrence. This implies that in three phase systems, even in case of a balanced short circuit, the DC components of different phases are not the same.

If the short circuit inductance is time dependent, the DC component could become larger than the AC component. In such cases, the current zero crossing will be missing over a few periods until the DC current becomes smaller than the AC current. Short circuit near a generator is a typical case where this phenomenon may be observed. The consequence for a switching device is much longer arcing periods and higher dissipation losses in the arc. This makes current interruption more demanding. The following simple example shows how a time dependent inductance may result in DC components larger than the steady state AC current. In this example, the source inductance suddenly increases from zero to its final value of L_1 at time t_1:

$$L(t) = \begin{cases} L_0 & t_0 < t < t_1 \\ L_0 + L_1 & t \geq t_1 \end{cases} \tag{3.6}$$

The governing equation of the circuit may be written as:

$$\begin{cases} L_0 \frac{di}{dt} + Ri = \sqrt{2}\,U \cos \omega t & t_0 < t < t_1 \\ (L_0 + L_1) \frac{di}{dt} + Ri = \sqrt{2}\,U \cos \omega t & t \geq t_1 \end{cases} \tag{3.7}$$

The initial condition for the first differential equation is as in the case considered previously, $i\,(t_0^+) = 0$. For the second differential equation, the initial condition can be derived based on the current continuity in the short circuit inductance L_0.

Based on these equations, the short circuit current flowing through the short circuit inductance of the network can be determined. The DC component is determined by the lower, initial inductance value L_0, and the steady state AC current is determined by the larger combined inductance of $L_0 + L_1$. Therefore, in this case the DC component of the short circuit current is larger than its AC component.

The differential equations governing the short circuit current near generator during faults are far more complex than this simplified example. The total short circuit inductance is gradually increasing with time. Consequently, in many real cases, a closed form solution does not exist, and the short circuit current has to be calculated by solving the governing differential equations numerically.

3.1.2 Recovery Voltage

3.1.2.1 Power Frequency Recovery Voltage

Not all phases of a three-phase breaker interrupt at the same instant. The phase in which the current is first interrupted is called the *first phase* or *first pole to clear*. Which phase this is, depends on the contacts separation and movement relative to the current waveforms. After the arc is extinguished at the current zero crossing, the voltages across each of the three poles of the breaker in general show an oscillatory behaviour during the transition from a value corresponding to the arc voltage to a value corresponding to the power frequency voltage.

The situation during interruption of an ungrounded three-phase short circuit is shown in Figs. 3.4, 3.5 and 3.6. The circuit diagram is in Fig. 3.4. The system is assumed purely inductive at short circuit. The current in each phase, the phase voltages prior to the interruption and the voltage across the contacts after the current

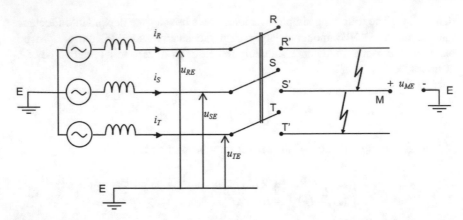

Fig. 3.4 Interruption of an ungrounded three-phase short circuit in a directly earthed grid. Stray capacitances are not included in the drawing. Load side impedances are disregarded

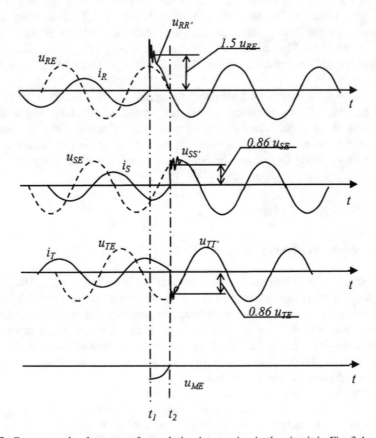

Fig. 3.5 Currents and voltage waveforms during interruption in the circuit in Fig. 3.4

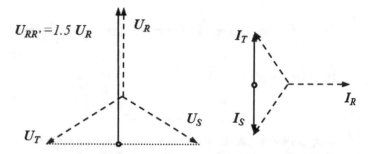

Fig. 3.6 Vector diagram for the case shown in Figs. 3.4 and 3.5. *Dashed lines* are used for $t < t_1$; *solid lines* for the period between t_1 and t_2 (i.e. when the arc is extinguished in phase R, but not in S and T)

has been interrupted are shown in Fig. 3.5, whereas Fig. 3.6 shows the corresponding vector diagrams for current and voltage.

The first phase to clear is here phase R, and the arc is extinguished at time t_1 in this phase. The voltage at the left hand side of the breaker in phase R then oscillates towards the phase voltage (i.e., the phase-to-ground voltage) of phase R. The right hand side voltage, R', oscillates towards a voltage level that is between the voltages in phases S and T. The arc is still burning in phases S and T and the voltages SS' and TT' are thus approximately zero.

From the voltage vector diagram, the power frequency recovery voltage in phase R is found to be 1.5 times the phase voltage of the system; U_p.

At t_2 phases S and T interrupt simultaneously, and the recovery voltage across each of the contact pairs becomes half the line voltage (i.e., half the line-to-line voltage), being $\sqrt{3}/2 \, U_p = 0.86 \, U_p$. From this follows that the interruption is easier in the last two poles to clear than in the first. Single-pole testing of switchgears is therefore acceptable when considering this switching duty.

The power frequency recovery voltage depends on the type of fault (single-phase, three-phase, short circuit, earth-fault etc.) and type of system grounding. If both the neutral point and the fault are directly in contact with ground, or grounded through small impedances, the power frequency recovery voltage becomes the same in all phases and is equal to U_p.

3.1.2.2 Transient Recovery Voltage in a Single-Phase Circuit

The voltage across the breaker contacts oscillates towards the power frequency recovery voltage via a transient period called the *transient recovery voltage* (*TRV*). As mentioned earlier, the degree of difficulty of the interruption process depends greatly on the shape and amplitude of the TRV.

The TRV will be examined in detail, starting with a single-phase system with a short circuit (relatively) close to the switchgear. The arc voltage is assumed small

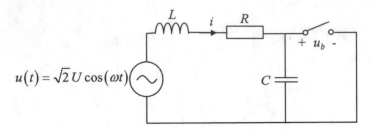

Fig. 3.7 Circuit diagram used to calculate the transient recovery voltage in a single-phase system after a current interruption due to a short circuit close to the switchgear

compared to the system voltage and is therefore neglected. Figure 3.7 shows the circuit diagram. U is the source voltage rms value, L the short circuit inductance in the system, R is a resistance representing the losses, and C the capacitance across the contacts.

The electric arc is assumed to be extinguished at the current zero crossing at $t = 0$, and no current flows through the switchgear for $t \geq 0$. The voltage that builds up across the breaker, u_b, shall be determined.

First, it is assumed that $R = 0$, and the following equations can be established:

$$u - L\frac{di}{dt} - u_b = 0 \tag{3.8}$$

$$i = C\frac{du_b}{dt} \tag{3.9}$$

By introducing (3.9) into (3.8), the following equation is found:

$$LC\frac{d^2u_b}{dt^2} + u_b = u \tag{3.10}$$

This is an inhomogeneous second order differential equation. The solution is the sum of the solutions to the corresponding homogenous differential equation (transient term) and one solution of the inhomogeneous equation (stationary term)

$$u_b = u_{bs} + u_{bt} \tag{3.11}$$

The transient solution is assumed to be of the form:

$$u_{bt} = A\,\sin\frac{t}{\sqrt{LC}} + B\,\cos\frac{t}{\sqrt{LC}} \tag{3.12}$$

where the initial conditions determine the constants A and B.

By considering the circuit as a voltage divider and using vector calculus, the rms value of the stationary solution is found as:

$$U_{bs} = \frac{\frac{1}{j\omega C}}{\frac{1}{j\omega C} + j\omega L} U = \frac{1}{1 - \omega^2 LC} U \qquad (3.13)$$

The stationary solution then becomes:

$$u_{bs} = \frac{\sqrt{2} U}{1 - \omega^2 LC} \cos \omega t \qquad (3.14)$$

and for the total solution:

$$u_b = \frac{\sqrt{2} U}{1 - \omega^2 LC} \cos \omega t + A \sin \frac{t}{\sqrt{LC}} + B \cos \frac{t}{\sqrt{LC}} \qquad (3.15)$$

By taking into account that there cannot be an instantaneous change in the voltage across a capacitance, nor a discontinuity in the current flowing through an inductance, the initial conditions can be found. Since the arc voltage is assumed zero and the current is interrupted at its zero crossing

$$u_b(0-) = u_b(0+) = 0 \qquad (3.16)$$

and

$$i(0-) = i(0+) = 0 \qquad (3.17)$$

Inserting (3.9) into (3.17) yields:

$$C \frac{du_b}{dt}\bigg|_{t=0} = 0 \qquad (3.18)$$

The initial conditions of (3.16) and (3.18) inserted into (3.15) give:

$$u_b(0) = \frac{\sqrt{2} U}{1 - \omega^2 LC} + B = 0 \qquad (3.19)$$

$$B = -\frac{\sqrt{2} U}{1 - \omega^2 LC} \qquad (3.20)$$

and

$$\frac{A}{\sqrt{LC}} = 0 \qquad (3.21)$$

$$A = 0 \qquad (3.22)$$

respectively. The solution then becomes:

$$u_b(t) = \frac{\sqrt{2}\,U}{1 - \omega^2 LC}\left[\cos \omega t - \cos \frac{t}{\sqrt{LC}}\right] \quad t \geq 0 \tag{3.23}$$

and consists of a power frequency stationary term and a transient term. The values of L and C are normally so that $\omega^2 LC \ll 1$, and $\frac{1}{\sqrt{LC}} \gg \omega$. The transient frequency is high compared to the system frequency, and the expression for the voltage across the breaker after interruption can thus be simplified to:

$$u_b(t) = \sqrt{2}\,U\left[1 - \cos \frac{t}{\sqrt{LC}}\right] \tag{3.24}$$

This waveform is shown in Fig. 3.8a.

Note that the maximum amplitude of the recovery voltage can reach nearly twice the amplitude of the source or system voltage (the dashed line).

The frequency of the transient recovery voltage is:

$$f_t = \frac{1}{2\pi\sqrt{LC}} \tag{3.25}$$

The higher the frequency, the steeper the recovery voltage and the more difficult the interruption of a given current becomes. The range of f_t is typically from a few hundred hertz and to a few kilohertz.

This type of transient recovery voltage, known as oscillatory transient recovery voltage, is normally observed in a wide variety of practical distribution network configurations. One example is the transformer-limited fault, where a fault occurs near one of the terminals of a circuit breaker connected directly to a transformer.

A slightly different case is when many overhead transmission lines are connected to one bus bar. If a terminal fault occurs near one of the circuit breakers of the overhead lines (see Fig. 3.9), the energy exchange between energy storing elements leads to generation of an over-damped voltage pulse. In the basic schematics of

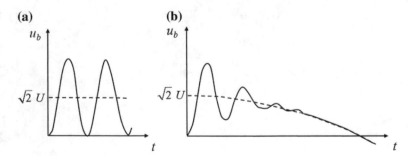

Fig. 3.8 Recovery voltage in a single-phase circuit after interruption of a short circuit close to the switchgear. If the transient frequency is very high compared to the power frequency and damping is disregarded the waveform is as in (**a**). The waveform becomes as in (**b**) if damping of the transient part is taken into account

Fig. 3.9, the energy-supplying network is replaced with its Thevenin equivalent circuit, where L_{sc} represents the equivalent short circuit inductance. Input impedance of the transmission lines seen from the busbar side Z_{in} is the same as their wave impedances Z_0, which is a pure resistance R_0, if one neglects the losses of the transmission lines. In addition, C_p represents the stray capacitance between busbar and ground. If the short circuit current has no DC component or the DC component has been already decayed, it can be expressed using a sine function. In case the short circuit current has a DC component, the situation could be slightly different.

The current is interrupted at its zero crossing ($t = 0$). The equivalent circuit for this basic network after current zero is shown in Fig. 3.10. Z_{eq} is the parallel

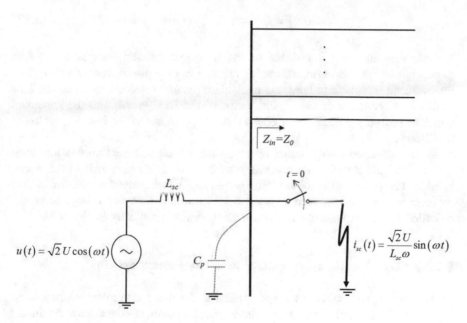

Fig. 3.9 Basic network for transient recovery voltage calculation in a system with many transmission lines connected to one busbar

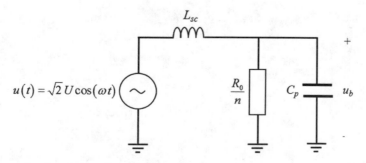

Fig. 3.10 Simplified equivalent circuit of the network shown Fig. 3.9

connection of all wave impedances, under assumption of no losses and the same wave impedances, it can be expressed as R_0/n, where n is the number of transmission lines.

The equivalent circuit of the network in this case is shown in Fig. 3.10. The differential equation of this circuit can be written as:

$$L_{sc}C_p \frac{d^2u_b}{dt^2} + \frac{nL_{sc}}{R_0}\frac{du_b}{dt} + u_b = \sqrt{2}\,U\,\cos(\omega t) \tag{3.26}$$

With the initial conditions:

$$u_b(t = 0^+) = 0 \quad and \quad \frac{du_b}{dt}(t = 0^+) = 0 \tag{3.27}$$

Note that the input impedance of the long transmission lines is their wave impedance. As the wave impedance is normally in the range of few hundred ohms for overhead transmission lines, the equivalent parallel resistor is quite small. This results in a reduction of the quality factor of the RLC circuit and the generated transient recovery voltage is, in contrast to the case shown in Fig. 3.8, no more oscillatory.

The overdamped voltage pulse, which propagates along other transmission lines, is reflected back at their ends, and reaches after a while the terminals of the circuit breaker. The superposition of the initially generated overdamped waveform and the reflected waveforms results in a very different transient recovery voltage form, the so-called double exponential waveform as shown schematically in Fig. 3.11.

3.1.2.3 Transient Recovery Voltage in a Three-Phase System

In this section, the recovery voltage applied to the first phase to clear when interrupting a short circuit current in a three-phase system is considered. As in the

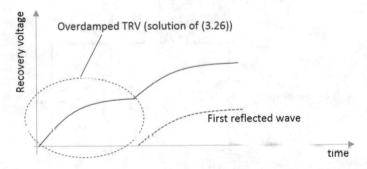

Fig. 3.11 Typical transient recovery voltage waveform in networks as shown in Fig. 3.9

single-phase case discussed in Sect. 3.1.2.2, the short circuit is assumed to occur close to the switchgear, a so-called *terminal short circuit*.

Two cases are treated here; a short circuit current in a solidly grounded system is interrupted with and without connection to the earth potential.

Figure 3.12 shows the circuit diagram for a three-phase short circuit without earth connection. C_p is the capacitance between the phases, and C_g between each phase and ground. Initially, the resistances in the circuit are disregarded. The arc in phase R is extinguished, so this phase is open. The arcs in phases S and T are still carrying current (through the arc).

The equivalent circuit as seen from the terminals of the first phase to clear, RR', is shown in Fig. 3.13. Only negligible current flows in the ground system, and the equivalent inductance L_e and equivalent capacitance C_e are found by short-circuiting all voltage sources and then determining the inductance and capacitance between R and R'. This gives:

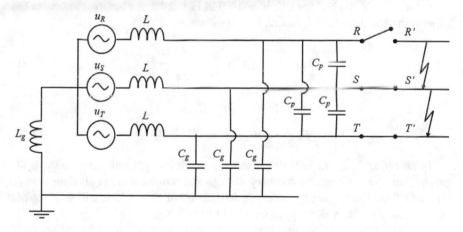

Fig. 3.12 Circuit diagram for interruption of an ungrounded terminal short circuit

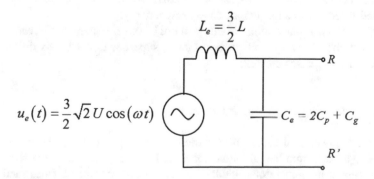

Fig. 3.13 Simplified equivalent circuit of Fig. 3.12

$$L_e = L + L||L = \frac{3}{2}L \tag{3.28}$$

and

$$C_e = \left(C_g + C_g||C_g\right)||\left(C_p||C_p\right) = 2C_p + \frac{2}{3}C_g \tag{3.29}$$

As in the case shown in Fig. 3.4, the equivalent voltage u_e is 1.5 u_p.

The equivalent circuit is analogous to the single-phase system in Fig. 3.7. Consequently, the solution is given as:

$$u_{RR'}(t) = \frac{\frac{3}{2}\sqrt{2}\,U_p}{1 - \omega^2 L_e C_e}\left[\cos \omega t - \cos \frac{t}{\sqrt{L_e C_e}}\right] \quad t \geq 0 \tag{3.30}$$

where C_e and L_e are given by (3.28) and (3.29).

A grounded three-phase short circuit corresponds to a direct grounding of the right hand side of the breaker (R′, S′ and T′) in Fig. 3.12. As before, the equivalent impedances are determined. In this case they are:

$$L_e = L + L||L||L_g = \frac{3}{2}L\frac{L_g + \frac{L}{3}}{L_g + \frac{L}{2}} \tag{3.31}$$

and

$$C_e = C_g||C_p||C_p = 2C_p + C_g \tag{3.32}$$

By assuming $L_g \gg L$, implying that only negligible current flows through the ground, the power frequency recovery voltage and thus also the equivalent voltage become 1.5 u_p. Consequently, the equivalent circuit of Fig. 3.13 is still valid, but L_e and C_e now take the values given by (3.31) and (3.32).

The characteristics of the recovery voltage waveform of Fig. 3.8 are thus valid for this case as well. In an inductively grounded system, i.e. with $L_g \gg L$, the transient recovery voltage frequency is fairly similar to the ungrounded and grounded three-phase short circuit cases, respectively.

Single- and two-phase short circuits will not be covered here. However, it can be shown that the recovery voltage becomes a superposition of two different frequencies in these cases as well.

3.1.2.4   Principle of Current Injection

In reality, a power grid cannot be represented as simple as it has been done here. Typically, a large number of inductances and capacitances are required to adequately represent generators, cables, transformers, overhead lines and other components. The equivalent circuits become so complex that computerized numerical calculations are needed to determine the recovery voltages with a reasonable accuracy.

Many software packages that can handle the mathematics of such problems are available, so the calculation itself may not be a problem. Finding an adequate circuit representation of the various system components may, however, prove to be a far more difficult task. For example, finding the inductances that represent a generator or transformer at high frequencies is not necessarily straightforward. It may also not be obvious whether they should be represented by a lumped inductance, or if a network of inductances and capacitances is required.

Instead of looking into electric network representation or numerical techniques for solving complex circuit equations, the current injection principle will here be examined. Current injection is a technique that can be used to determine the recovery voltage both through measurements and by analytical calculations.

The current injection principle is based on the following observations: With regard to currents and voltages in a system, current interruption can be equalized by injecting a current across the contacts of the switching device from the moment of interruption and onwards, i.e. for $t > 0$. The injected current is equal in magnitude to the current that would have passed if there had been no interruption, but has opposite polarity. This is illustrated in Fig. 3.14. The source side is represented by the Thévenin equivalents u_T and Z_T. The two ideal current sources in parallel (they have infinite internal impedance) represent the switching device.

According to the superposition principle, the currents and voltages in a linear system can be found by calculating the currents and voltages from each source independently while others are de-activated, (i.e. voltage sources are short-circuited and current sources are open) and then adding the contributions from all sources.

There are two contributions to the voltage across the contacts in Fig. 3.14, when the power switching device is in closed position (for $t < 0$). The voltage source

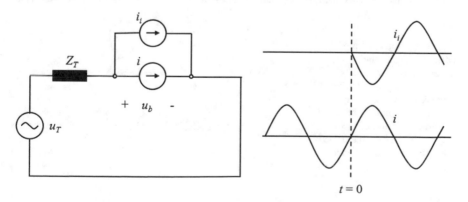

Fig. 3.14 The current injection principle. Interruption takes place at $t = 0$. The *curves* on the right are the currents in the sources across the switchgear. The source i_i does not carry current for $t < 0$, for $t > 0$ it carries the current $i_i = -i$

contributes with u_T, while the current source i gives a contribution equal to $-u_T$. These summed gives $u_b = 0$, corresponding to closed contacts. The current source i_i does not contribute for $t < 0$ and, accordingly, no voltages in the circuit are generated by this source.

The current source i_i starts supplying current at $t = 0$. The sum of the currents through the power switching device is zero as $i_i = -i$, which corresponds to open contacts. The voltage across the contacts is calculated by considering all the three active sources (i, i_i and u_T). The voltage contributions from the stationary sources i and u_T cancel (as for $t < 0$), and the only contribution left is from i_i (with i open and u_T short-circuited). This is shown in Fig. 3.15.

The recovery voltage across the power switching device is thus entirely due to the injected current i_i, and only this contribution needs to be considered for determining the recovery voltage.

This approach may also be used in practical measurements for finding the recovery voltage across a switching device in a system, see Fig. 3.15

All voltage sources in the system being examined are short-circuited and a relatively low amplitude sinusoidal current is injected into the breaker. The transient recovery voltage across the contacts is at the same time measured with an oscilloscope. Then, the recorded voltage is up-scaled by using the ratio between the actual short circuit current and the injected current, and the actual recovery voltage, which the breaker may be exposed to, is found from the measured value (Fig. 3.16).

Normally, the recovery voltage immediately after current zero crossing is the most interesting part. As a sine wave is approximately linear right after current zero crossing ($\sin x \approx x$ for small x), it is usually sufficient to apply a linearly increasing current to represent the sine wave. The current source must be ideal, i.e. the impedances in the considered grid should not influence the injected current.

There are a number of obvious weaknesses to this method. The system in which the switchgear is operating must exist and be available for measurements, i.e. the technique cannot be used when planning new installations. Moreover, it may be very hard to get an entire system out of service to do measurements. However, in

Fig. 3.15 Application of current injection and superposition for calculation of the recovery voltage across a breaker. The current i_i has the same magnitude, but opposite direction of the stationary current i that passed through the closed contact

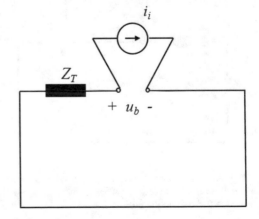

Fig. 3.16 Current injection across the contacts used to measure the recovery voltage during current interruption

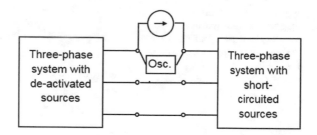

many cases including just a part of the grid and considering the rest as an ideal system with zero impedance will give adequate results.

There may also be difficulties due to rather large noise levels in real systems. This makes it necessary to use relatively high signal levels. However, despite all its limitations, the method serves as a relatively simple way of measuring recovery voltages.

3.1.2.5 Short Line Fault (Kilometric Fault)

The current injection principle and superposition may also be used to *calculate* recovery voltages. An important switching duty; interruption of short circuit currents when the fault is located a short distance away from the power switching device, will now be looked into by means of the current injection method.

In the calculations of recovery voltage in Sects. 3.1.2.2 and 3.1.2.3, it was assumed that the short circuit occurs close to the switchgear (terminal short circuit). Since the impedance on the load side of the switch in this case is negligible, the resulting short circuit current has the highest magnitude possible in the system considered. However, when also taking the amplitude and steepness of the recovery voltage into consideration, the terminal fault may still not be the most demanding switching duty for the breaker.

Generally, short circuits often occur some distance from the switchgear, for example between the phases of an overhead transmission line. Due to the steepness of the recovery voltage immediately after current zero crossing, short circuits that occur between a few hundred meters and a few kilometres away from the switching device turn out to be particularly difficult to interrupt. This type of fault is called *short line fault* or *kilometric fault*. When considering this switching duty, the impedances on the load side must also be included in the calculations. Only the single-phase case will be considered here; Fig. 3.17 shows the circuit diagram.

The short circuit inductance of the system is L_s. The inductance between the switching device and the short circuit is Ll, where L is the self-inductance per unit length of the transmission line and l the distance between the switch and the location of the short circuit.

The current injection principle is employed to determine the recovery voltage. A current is injected across the contacts from $t = 0$. The current has the same

Fig. 3.17 The circuit
diagram used for transient
recovery voltage calculations
during interrupting of a short
line fault

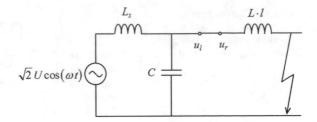

magnitude, but opposite polarity as the stationary current that were flowing through
the switch before interruption. By ignoring the system capacitance to earth, the
injected current becomes:

$$i_i(t) = -i(t) = -\frac{\sqrt{2}U}{\omega\,(L_s + Ll)}\sin\omega t \tag{3.33}$$

The voltage on the left side of the switchgear due to the injected current, u_{li}, is
found by short-circuiting the voltage source in the circuit in Fig. 3.17. The resulting
circuit diagram is shown in Fig. 3.18.

By using Kirchhoff's current law at the node to the left of the switchgear, the
following equation can be established:

$$C\frac{du_{li}}{dt} + i_{Ls} + i_i = 0 \tag{3.34}$$

In addition,

$$u_{li} = L_s\frac{di_{Ls}}{dt} \tag{3.35}$$

The differential equation for u_{li} is found by differentiating (3.34) and (3.35), and
combining with (3.35).

$$\frac{d^2u_{li}}{dt^2} + \frac{1}{L_sC}u_{li} = \frac{\sqrt{2}U}{C\,(L_s + Ll)}\cos\omega t \tag{3.36}$$

Fig. 3.18 Circuit diagram
used for calculating the
voltages due to the injected
current

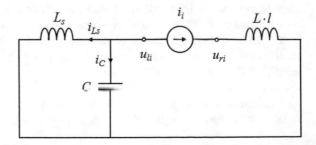

The solution to this equation is a sum of a transient and a stationary term. The transient term is on the form:

$$u_{lit} = A \, \sin \frac{t}{\sqrt{L_s C}} + B \, \cos \frac{t}{\sqrt{L_s C}} \tag{3.37}$$

where the initial conditions determine the constants A and B.

The stationary solution is found by assuming that the power frequency current flowing through the capacitance C is negligible:

$$u_{lis}(t) = -i_i(t)\omega L_s = \frac{\sqrt{2} U L_s}{L_s + Ll} \cos \omega t \tag{3.38}$$

The general solution differential equation determining the left side voltage due to the injected current then becomes

$$u_{li} = u_{lit} + u_{lis} = A \, \sin \frac{t}{\sqrt{L_s C}} + B \, \cos \frac{t}{\sqrt{L_s C}} + \frac{\sqrt{2} U L_s}{L_s + Ll} \cos \omega t \tag{3.39}$$

The initial conditions can be found by considering that: (i) there cannot be a discontinuity in the voltage across a capacitance, and, (ii) there cannot be a discontinuity in the current passing through an inductance. At $t = 0$, when the current injection is initiated, the voltage drop across the capacitance C is as large as the voltage over the short transmission line:

$$u_{li}(0^+) = u_{li}(0^-) = \frac{\sqrt{2} U Ll}{(Ll + L_s)} \tag{3.40}$$

Inserted into (3.39), this gives:

$$B = -\frac{\sqrt{2} U L_s}{L_s + Ll} + \frac{\sqrt{2} \, U \, Ll}{(Ll + L_s)} \tag{3.41}$$

At $t = 0$, no current passes through the inductances or any other part of the circuit.

$$i_{Ls}(0+) \ = \ i_C(0+) \ = \ 0 \tag{3.42}$$

This implies that:

$$\frac{du_{li}(0^+)}{dt} \ = \ 0 \tag{3.43}$$

Inserted into (3.39) gives:

$$A = 0 \tag{3.44}$$

The voltage on the left side of the switch due to the injected current thus becomes:

$$u_{li} = \frac{\sqrt{2}UL_s}{L_s + Ll}\left[\cos\omega t - \cos\frac{t}{\sqrt{L_sC}}\right] + \frac{\sqrt{2}\,U\,Ll}{(Ll+L_s)}\cos\frac{t}{\sqrt{L_sC}} \tag{3.45}$$

The voltage on the right side of the switchgear caused by the injected current, u_{ri}, is found by considering the overhead line between the breaker and the location of the short circuit as a wave impedance Z with travelling time τ. The overhead line is short-circuited (the refection coefficient $\rho = -1$) at the location of the fault, and connected to the current source (large impedance, reflection coefficient $\rho = 1$) at the switchgear.

The current i_i is injected across the contacts, and the current sine is approximated with a ramp function as only the time interval in the very beginning of the recovery voltage is considered, i.e.

$$i_i \approx \left.\frac{di_i}{dt}\right|_{t=0+} t = -\left.\frac{di}{dt}\right|_{t=0+} t \tag{3.46}$$

This current leads to a voltage wave:

$$Z\,i_i \approx -Z\frac{di}{dt}t = -Z\frac{\sqrt{2}U\omega}{\omega\,(L_s + Ll)}t = -Z\frac{\sqrt{2}U}{(L_s + Ll)}t \tag{3.47}$$

travelling back and forth between the location of the short circuit and the breaker, as shown in the reflection diagram in Fig. 3.19.

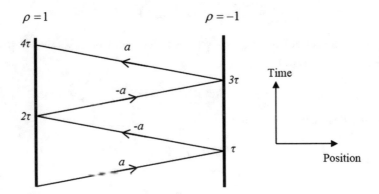

Fig. 3.19 Reflection diagram for the travelling wave on the overhead line between the switchgear ($\rho = 1$), and the location of short circuit ($\rho = -1$)

The voltage on the right side of the breaker caused by the injected current can be expressed as:

$$u_{ri} = \frac{\sqrt{2}U\,Ll}{(L_s + Ll)} - Z\frac{\sqrt{2}U}{(L_s + Ll)}\left[t - 2(t - 2\tau)H(t - 2\tau) + 2(t - 4\tau)H(t - 4\tau)\right.$$
$$\left. - 2(t - 6\tau)H(t - 6\tau) + \ldots\right] \quad \text{where } H(t) = \begin{bmatrix} 0 & \text{for } t < 0 \\ 1 & \text{for } t \geq 0 \end{bmatrix} \tag{3.48}$$

It must be considered that the transmission line at the time of current interruption at current zero crossing is charged and has an initial voltage of:

$$u_{ri} = \frac{\sqrt{2}U\,Ll}{(L_s + Ll)} \tag{3.49}$$

The recovery voltage across the breaker due to the injected current is the difference between the voltages to the left and to the right of the switchgear

$$u_b = u_{li} - u_{ri} \tag{3.50}$$

When (3.45) and (3.48) are inserted into (3.50), this expression becomes:

$$u_b = \frac{\sqrt{2}UL_s}{L_s + Ll}\left[\cos\omega t - \cos\frac{t}{\sqrt{L_sC}} + \frac{Z}{L_s}[t - 2(t - 2\tau)H(t - 2\tau) + \ldots]\right] + \frac{\sqrt{2}\,U\,Ll}{(Ll + L_s)}\left(\cos\frac{t}{\sqrt{L_sC}} - 1\right) \tag{3.51}$$

The transient recovery voltage is a superposition of voltages with three different frequencies: a power frequency part (50 or 60 Hz), a part with a frequency determined by the reactance at the source side (typically a few kilohertz) and a part due to the travelling waves on the load side (typically a few hundred kilohertz). The source side transient recovery voltage u_{li}, which is a combination of two oscillatory voltages with different frequencies, is shown in Fig. 3.20. Figure 3.21 shows both source side and load side transient recovery voltages and the resulted transient recovery voltage applied to the switching device.

The lowest frequency contributions cancel for $t \ll \sqrt{L_sC}$, and the travelling wave term thus dominates. The first maximum of this term is:

$$u_{bmax} = Z\frac{\sqrt{2}U}{L_s + Ll}2\tau \tag{3.52}$$

for $t = 2\tau$. If the distance between the switchgear and the fault is 1 km, the maximum transient recovery voltage occurs less than 10 μs after current zero crossing. A voltage of this steepness such a short a time after the arc is extinguished leads to a considerable stress on the switching gap and the switchgear has to be carefully designed to avoid thermal re-ignition.

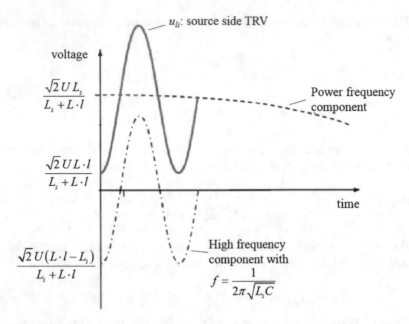

Fig. 3.20 Source side transient recovery voltage in case of a short line fault

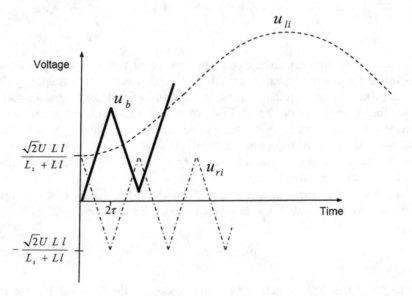

Fig. 3.21 Source side, load side and total transient recovery voltage in case of a short line fault

In reality, the situation is more complex. Among the other things that also should be taken into account are:

- Real systems are three-phase systems.
- The travelling waves are damped.
- The source side cannot be represented by a single inductance.
- Equipment installed on the line side may interact and causing the wave impedance description to be flawed.

However, the characteristic features causing the difficulties experienced when interrupting a short line fault are evident in this simple description. Depending on the current interruption technique and design of the breaker, the most demanding case may occur at different fault current levels. Therefore, as described in Sect. 3.1.3, different fault current levels (e.g. 75% of the short circuit current or 90% of the short circuit current) are used in tests for verifying the switching performance of circuit breakers in the case of short line faults.

Although the short circuit current amplitude is smaller in short-line faults compared to the terminal fault, the rate of rise of recovery voltage is higher. For switching devices with long recovery times, the very fast increase of recovery voltage applied to the switching gap may result in a thermal re-ignition. As shown in (3.47), the rate of rise of recovery voltage of the saw-tooth like waveform on the load side of the switching device is dependent on the wave impedance of the transmission line. Therefore, the transmission systems with higher wave impedances (e.g. overhead lines) may be more critical from the perspective of short line fault compared to those with smaller wave impedances (e.g. power cables).

3.1.2.6 Out of Phase Interruption

Circuit breakers that connect power generating plants to the grid, or link different parts of a greater power system together, must be capable to interrupt fault currents even though the two system parts are not running synchronously, i.e. the voltages in the two parts are out of phase. Power system stability problems may lead to such a situation. Normally, fast relaying systems will trip the breakers while the phase shift is small. However, if the primary relaying system malfunctions, some time may pass and the phase shift may be substantial by the time the circuit breaker receives the command signal to open.

Out of phase interruption is probably more likely to occur under the interruption following an unsuccessful automatic re-closure. Automatic re-closure is quite often used in modern relaying systems in order to clear intermittent faults without interrupting the power supply. It essentially means that when a fault signal is received, the circuit breaker will open, but then immediately close after a short delay. If the disturbance is of an intermittent nature, e.g. a lightning stroke on an overhead line, the service is restored within a few hundred milliseconds. If, on the other hand a permanent failure that damages equipment has occurred, the circuit breaker immediately interrupts the current again, and remains in open position until the failure is repaired.

During the short un-energized interval between the two interruptions the rotational speed of the generators in the two system parts connected by the breaker may drift independently. The generators in the part that looses load accelerate, and the generators in the part with increased load retard, and a phase shift thereby develops between the voltages of the two systems. If the phase shift has become too large by the time the automatic re-closure occurs, the two system parts continue to run asynchronously. This is a dynamic stability failure, and the interruption that follows may take place at any phase angle difference between the two system parts, including a full 180° phase opposition. In a directly grounded system the power frequency recovery voltage becomes about twice the phase voltage, see Fig. 3.22. If the switchgear is located at the end of a long transmission line, the Ferranti effect may cause the voltage to become even higher.

The lower part of Fig. 3.22 illustrates the case with a double ground fault, one in each part of the system, in phases R and S respectively. Moreover, it assumed that a

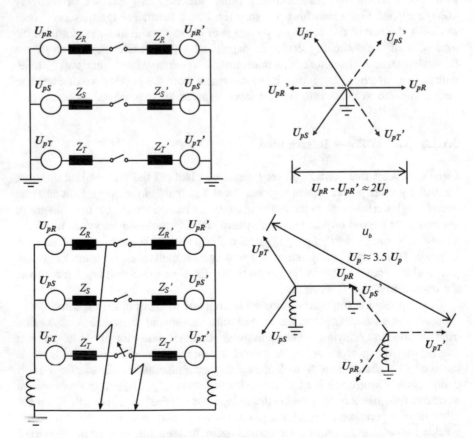

Fig. 3.22 Power frequency recovery voltage during interruption at phase opposition (*upper part*), and with ground faults on both sides of the switchgear and a very disadvantageous phase shift between the systems (*lower part*)

phase shift of 120° has occurred between the two systems. The switchgear has interrupted the current in phases R and S, and the vector diagram shows that the power frequency recovery voltage in phase T is close to $3.5U_p$.

However, due to the line impedances, the current amplitudes under out of phase interruptions are always smaller than in the terminal short circuit case. The current depends on the system impedances and the location of the breaker.

The steepness of the recovery voltage during out of phase interruption is normally modest. This is explained by Fig. 3.23 for a breaker located at the end of a long transmission line.

It has been shown (see (3.12) and Fig. 3.8) that the transient recovery voltage frequency across a breaker having an LC-circuit on the source side is:

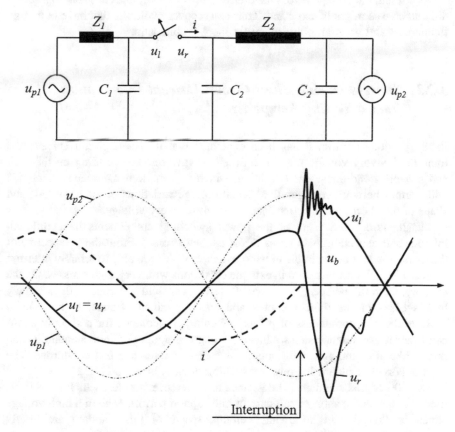

Fig. 3.23 The circuit diagram (*upper part*) and the corresponding current and voltage waveforms (*lower part*) during interruption under full phase opposition (180°). The breaker is assumed located close to the source u_{p1} and there is a long line between the breaker and u_{p2}. This means that $Z_1 < Z_2$ and $C_1 < C_2$, the transmission line to the right of the breaker is represented by its π-equivalent

$$f_t = \frac{1}{2\pi\sqrt{LC}}$$

(3.53)

In the circuit in Fig. 3.23, the impedances on the left side of the switchgear are smaller than on the right side. Consequently, after interruption u_l has a higher frequency than u_r.

The change in amplitudes for the voltages u_l and u_r are also influenced by the impedances. Since the impedances between the breaker and the source are smaller on the left side, u_l is far less altered than u_r. Consequently, the high frequency transients to the left of the switchgear have relatively low amplitudes. The situation is opposite on the right side. Here, the changes in amplitude are greater, but the frequencies of the transients are relatively low.

As a result, the voltage across the breaker, $u_b = u_l - u_r$, does not lead to a such a difficult case as might be expected. Transient recovery voltages that have both high frequencies and great amplitudes are far more problematic.

3.1.3 Stresses During Short Circuit Current Interruption from a Testing Perspective

In the previous section, it has been explained that the short circuit current and transient recovery voltage may have different waveforms, depending on the fault and network conditions. Fault location, which is a random parameter, combined with various network configurations result in large variations of the amplitude and shape of the short circuit current and transient recovery voltage.

Manufacturers and users of the power switching components have different interests and expectations in terms of the applied stresses. From the perspective of the users, it would be desirable to have switches, which are able to reliably interrupt all possible fault currents even in extreme conditions with very high stresses. On the contrary, from the perspective of manufacturers, capability to cope with extremely high stresses means more complex and costly products. Hence, to be able to evaluate the appropriateness of power switching components for a specific application, a set of reference stresses have to be agreed upon. Different standardisation bodies like IEC and ANSI formulate these sets of reference test conditions. The most important standard for power switching devices is the IEC 62271.

Part 100 of this standard [3] describes the reference stresses, which have to be used to evaluate the short circuit current interruption performance of a high voltage circuit breaker. In the following, certain aspects of this standard are briefly reviewed.

The short circuit current is assumed to be sinusoidal with or without a DC component, i.e. symmetrical or asymmetrical in the range of 10–100% of the rated short circuit current of a power switching device. This is to cover all scenarios a

switching device could face in a real power network. It is preferable to test three
phase circuit breakers under three phase conditions, but single-phase tests are also
allowed. The DC component of the short circuit current is limited to 20% of the
rated short circuit current in case of symmetrical test duties for not self-triggered
switching devices; this is because the DC component has been decayed before the
contacts of the switching device separate. If the switching device is self-triggered or
is supposed to act very fast (as generator circuit breakers), the DC component could
be much larger. The impact of having a DC component is larger short circuit current
amplitudes and longer maximum arcing times.

Transient recovery voltage, as explained earlier, is very much dependent on the
network configuration. In the standards, there is a differentiation based on the rated
voltage of the switching device. For rated voltages less than 100 kV, the so-called
two parameter TRV is proposed. In Fig. 3.24, two envelopes of a standard two
parameter TRV are shown. For rated voltages higher than 100 kV, the so-called
four parameter TRV is used, see Fig. 3.25. This corresponds to the waveforms
applied to the switching device in networks with many transmission lines connected
to one busbar, see Fig. 3.11.

The TRVs of Figs. 3.24 and 3.25 are applied to the switching device when its
capability to interrupt terminal faults is to be tested. As explained in Sect. 3.1.2.5, if

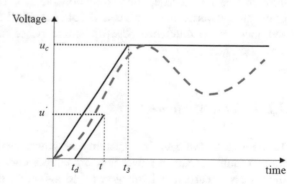

Fig. 3.24 Two parameter
transient recovery voltage
representation

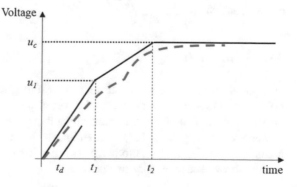

Fig. 3.25 Four parameter
representation of transient
recovery voltage

the short circuit fault is not directly at the terminal of switching device, a much steeper transient recovery voltage is generated. The short line fault is only critical if the wave impedance of the transmission system is high, because the rate of rise of recovery voltage is proportional to the wave impedance. This is the case in high voltage networks where usually overhead transmission lines are used. With this background, the short line fault testing is according to the standards only required for switchgear with rated voltages above 100 kV. As explained earlier, changing the length of the line on the load side of the switching device, has an impact on the current amplitude as well as on the rate of rise of recovery voltage. Depending on the current interruption characteristics of the breaker, a certain line length is the most difficult case. To ensure a satisfactory ability to handle short line faults, the standard prescribes tests with different short circuit levels, from 60 to 90% of the rated short circuit currents (corresponding to different line lengths). Moreover, the standards allow combining short line fault tests with other breaking test duties, if a certain initial transient recovery voltage is applied.

If current interruption tests are performed single phase, an additional correction factor for the applied TRV, due to different neutral grounding situations, has to be taken into account. In Sect. 3.1.2.3, it has been shown that the TRV of the first pole to clear is 1.5 times of the phase voltage in networks with an insulated neutral. Analogously, it can be shown that in systems where the neutral point is grounded through a low impedance, this correction factor is between 1 and 1.5. In solidly grounded networks, the currents and voltages of all three phases are independent, and there is no difference between single phase and three-phase TRV, so the correction factor is 1.

3.1.4 Test Methods

To be able to evaluate the capability of power switching components for interrupting fault currents in power networks, laboratory testing is required. The specified stresses (current and voltage waveforms) are applied to the switching devices, but there are several different approaches that can be used to achieve the prescribed testing conditions.

3.1.4.1 Direct Tests

One possibility is to use a very strong power source with a voltage level of the rated voltage of the switching device to be tested and to provide the necessary short circuit current. This method is normally referred to as *direct testing*. The needed power source may be a large generator, which can operate under short circuit condition, see Fig. 3.26a. Another possibility is to obtain the required power from grid as shown schematically in Fig. 3.26b.

(a)

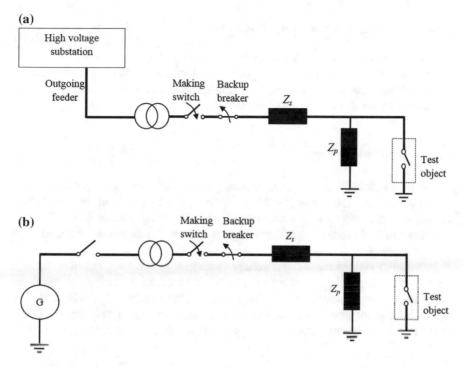

Fig. 3.26 Schematics of test circuits with power source as **a** generator **b** power network

In these circuits, Z_s is the series impedance used to adjust the short circuit current level, which is usually a current limiting inductor. Z_p is the parallel impedance to the test object, which is in many cases a series combination of a capacitor and a resistor. The main purpose of inserting a parallel branch to the test object is to shape the applied TRV after interruption of the current.

The test is started by closing the making switch of the test circuit and then opening of the test object as current flows. A circuit breaker (backup breaker) is normally placed in series and used to ensure reliable current interruption in case of any failure of the test object.

Direct testing of power switching components can be easily performed in a three phase setup. Then the circuits of Fig. 3.26 has to be realized with three series and three parallel impedances. Obviously, a three phase transformer and generator are also required.

The main challenge associated with direct testing of switchgear is the rating of the power source. As both short circuit current and the necessary transient recovery voltage are supplied by one single power source, it has to be very powerful. Consider for example the case where the current interruption capability of a 420 kV circuit breaker with rated short circuit current of 50 kA has to be tested. In this case, the necessary three-phase short circuit power is:

$$P_{SC} = \sqrt{3} \times 50 \times 10^3 \times 420 \times 10^3 \approx 3.64 \times 10^{10} \text{ W}$$

This is a huge number, greater than the combined installed power generation in a country like Norway or the greatest power plant in the world (The Three Gorges dam in China). Obviously, testing of transmission level circuit breakers is almost impossible by using the direct testing method.

3.1.4.2 Synthetic Testing

As the high voltage and high current during a current interruption process do not appear at the same time, it is possible to divide the test of a circuit breaker in two distinct regions, namely high current and high voltage regions, see Fig. 3.27. Test arrangements and procedures taking advantage of this are commonly referred to as *synthetic testing*.

In the high current region, the voltage necessary to drive high current through the switching device is almost zero (when the contacts are closed) or equal to the arc voltage, which is very much smaller than the system voltage. In the high voltage region, the current flowing through the power switch is zero or very small, at least in case of a successful current interruption. This implies that application of two

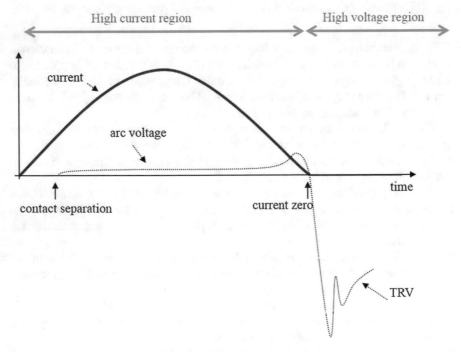

Fig. 3.27 High current and high voltage regions during short circuit current interruption

separate and rather modestly rated power sources can be used to supply energy to
the under test power switch at different times.

Depending on how these two power sources are coupled to generate the electrical stresses on the switchgear, different test circuit configurations exist. Two
well-known methods are the parallel and series current injection, shown schematically in Fig. 3.28. In all configurations used to realize synthetic test circuits, an
appropriate mechanism to synchronize the high current and high voltage sources is
necessary.

The high current source can be a large generator or a grid-connected circuit as
shown in Fig. 3.26. A charged capacitor bank in connection with an inductor and
with the resonance frequency near the power network frequency can also be
applied. The high voltage source is usually a high voltage capacitor bank. The
TRV-shaping circuit can be as simple as a capacitor connected in parallel to the test
object or a combination of many components, depending on the desired TRV
waveform. The most complicated TRV-shaping circuits are related to short-line
fault tests, where a lumped model for short lines is built connecting several different
passive components.

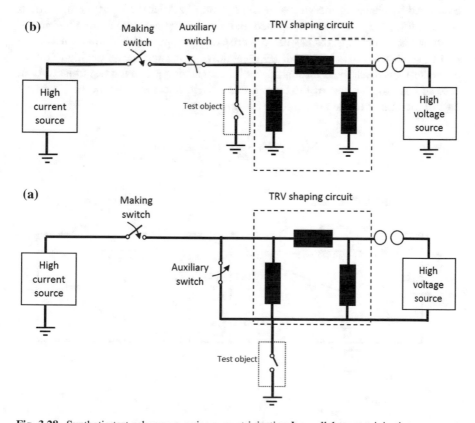

Fig. 3.28 Synthetic test schemes **a** series current injection **b** parallel current injection

The working principle of these synthetic test methods can be explained by considering the simple circuit shown in Fig. 3.29. Here, both high current and high voltage sources are realized in a parallel current injection scheme using capacitor banks.

Switch 1 is the auxiliary switch, switch 2 is the making switch and the test object (TCB) refers to the circuit breaker, which is to be tested. The current and voltage, which the test object is subjected to, are measured by means of a shunt resistor and a voltage divider. The TRV shaping circuit is the simplest possible one, consisting of a capacitor connected in parallel to the test object.

At the start of the test, switch 1 and the test object are closed and switch 2 is open. The test begins with closing of the making switch. Closing of the making switch results in discharge of capacitor C_1 through inductance L_1 and in generation of a short circuit current:

$$i_1(t) = U_1 \sqrt{\frac{C_1}{L_1}} \sin\left(\frac{t}{\sqrt{L_1 C_1}}\right) \tag{3.54}$$

The amplitude and frequency of the generated short circuit current can be simply controlled by changing the values of C_1, L_1 and U_1. After the short circuit current starts to flow, the test object is opened and a switching arc ignites. Just before the short circuit current approaches its zero crossing, the spark gap of the high voltage part of the test circuit is triggered. By breakdown of the spark gap, the high voltage capacitor C_2 is discharged through L_2 and the switching arc in the test object. As in case of high current part of the circuit, this current injected from the high voltage part of the circuit to the switching arc is given as:

$$i_2(t) = U_2 \sqrt{\frac{C_2}{L_2}} \sin\left(\frac{t - t_{sg}}{\sqrt{L_2 C_2}}\right) \tag{3.55}$$

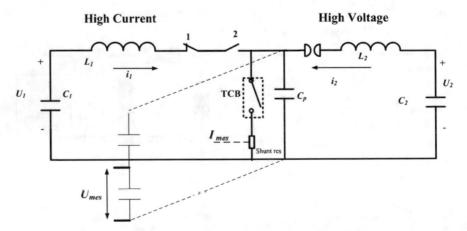

Fig. 3.29 Simplified synthetic test circuit based on parallel current injection method

Fig. 3.30 Equivalent circuit of Fig. 3.29 just before its current zero crossing

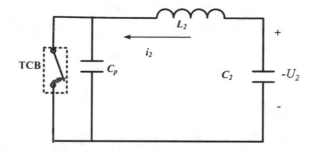

So, from the time t_{sg} (the time of breakdown of the spark gap) and onwards, the sum of the currents i_1 and i_2 flows through the arc. Consequently, the zero crossing of the current flowing through the test object is shifted, while the zero crossing of the current flowing through the auxiliary switch (i_1) comes earlier. A successful current interruption in the auxiliary switch splits the high current and high voltage parts of the synthetic test circuit. The equivalent of the test circuit just before the zero crossing of $i_1 + i_2$ is shown in Fig. 3.30.

Note that the voltage across capacitor C_2 at the time of current zero crossing is nearly $-U_2$, if the losses are negligible. Just after the current zero, the test object opens and acts like an ideal switch. As a result, between two capacitors with different initial voltages (C_p with an initial voltage almost zero and C_2 with an initial voltage $-U_2$), an oscillation takes place. In this way, an oscillatory voltage is generated over C_p and the test object, giving a TRV as:

$$u_{TRV}(t) = -U_2 \frac{C_2}{C_2 + C_p} \left[1 - \cos\left(\frac{t - t_{cz}}{\sqrt{L_2 \frac{C_2 C_p}{C_2 + C_p}}}\right) \right] \quad t > t_{cz} \qquad (3.56)$$

In real test circuits, there are some losses due to the series resistance of L_2 and therefore the transient recovery voltage becomes a damped oscillatory waveform. This form represents very well the two parameter TRV described in Sect. 3.1.3. It is possible to increase the initial rate of rise of the recovery voltage by inserting a series resistance to the parallel capacitance C_p.

The making switch, the auxiliary switch as well as the test object are all mechanically operating circuit breakers. Therefore, to be able to open or close these switches at a specific time, open/close commands have to be sent much earlier. To minimize the energy content and size of the high voltage capacitors, the frequency of the injected current is as high as possible, normally up to 1000 Hz. Thus, it is crucial for the correct functioning of the synthetic test circuit to trigger the spark gaps very precisely. This implies a very good synchronization between the high current and injected current. Figure 3.31 shows the sequential stages of the operation of a parallel current injection synthetic test circuit.

The test starts by the sending a closing command to the making switch (switch 2 in Fig. 3.29). Depending on the mechanical opening and closing times, the opening

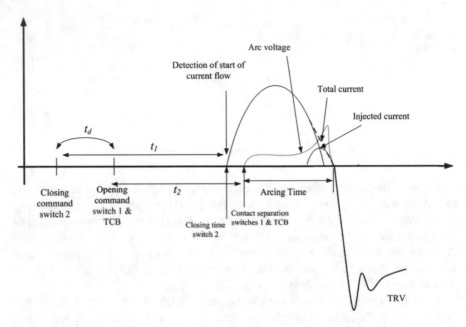

Fig. 3.31 Sequential stages of a synthetic test operation

command is sent after a delay of t_d to the auxiliary switch and to the test object. The aim is to separate the contacts of those switches when the current flows.

By separation of the contacts in the test object, a switching arc is initiated. This arc burns between contacts of the test object until current zero crossing. The start of current flow is detected and after a certain time t_{d2}, the spark gap is triggered. This results in flow of an additional current (injected current), see (3.55) as discussed earlier. At the zero crossing of the high current flowing through the auxiliary switch, it is interrupted and the high current part of the test circuit is isolated from the other parts. At the zero crossing of the total current flowing through the test object, the current is interrupted and a transient recovery voltage, like as (3.56), builds up across the open contacts of the test object. In case of a successful current interruption, the under test switching device can withstand this voltage. If, in contrast the stresses cause a thermal or dielectric re-ignition, the test is considered failed.

3.2 Closing Under Fault Condition (Making of Short Circuit)

Connecting and disconnecting different parts of a power network is, besides fault current interruption, one of the most important functions of switchgears. If the part of the network, which is to be energized, is faulty, closure of the switching component results in a short circuit current flow through the switch during the closing (making) operation.

At first glance, it seems easy for a switching device to pass the short circuit current during a making operation, because all power switching devices have to be designed in such a way that they are able to carry the short circuit current for up to a few seconds while in the closed position.

However, the current flow does not start as the contacts touch but several milliseconds earlier because the insulating medium between the contacts suffers a dielectric breakdown as shown in Fig. 3.32. This phenomenon is called a *pre-strike*. The electrical field in the gap between the contacts increases as the contacts approach each other. When the field exceeds the dielectric strength of the insulating medium, the gap breaks down and a switching arc is initiated.

Flow of the short circuit current through this arc results in power dissipation, which is partly absorbed by the contact surfaces and partly contributes to increasing the temperature in the arcing medium. If the dissipated energy is large, excessive melting of the contact surfaces occurs. Mating contacts with molten surfaces may cause welding. This may prevent the switching device to appropriately respond to the next opening command, if its operating mechanism does not provide an opening force sufficient to break the welded points. The dissipated energy can be calculated as follows:

$$E_{dissipation} = \int\limits_{t_0}^{t_{touch}} i_{sc}(t) \cdot u_{arc}(t)dt \qquad (3.57)$$

$E_{dissipation}$ is the dissipated energy in the pre-strike during making operation, i_{sc} and u_{arc} are the short circuit current and arc voltage, respectively. t_0 is the moment

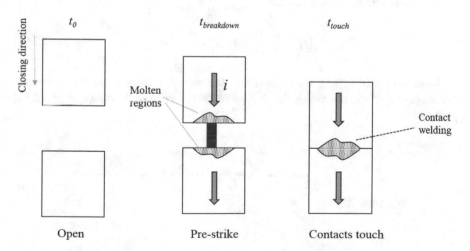

Fig. 3.32 Different stages during making operation of a power switching device

when the breakdown occurs, and t_{touch} is the moment when the contacts touch mechanically. Hence, a short circuit arc is burning in this time interval.

As explained in Sect. 3.1.1.2, the short circuit current contains a DC component and therefore its peak value can be much higher than for a pure AC short circuit current. The arc voltage is very dependent on the interrupting medium, but as explained in Chap. 2, even in case of very short arc lengths, there are significant voltage drops near the electrodes. The associated heat dissipation contributes to the energy flux to the surface of the contacts and their melting.

Making under short circuit conditions may occur in both circuit breakers and load break switches. Circuit breakers are designed to handle the high arcing energies occurring during interruption of short circuit currents. Hence, the energy dissipation during a pre-strike under making under short circuit conditions is usually not critical for a circuit breaker. Moreover, the opening and closing velocities of circuit breakers are rather high, resulting in a short arcing time.

For load break switches, in contrast, the making operations are among the most severe stresses, as these devices are designed to be able to cope only with the arc energies during load current interruption, and the closing or opening velocities are lower than for circuit breakers. The longer the arcing time by making operation, the higher the dissipated energy. This fact can be demonstrated considering the following example, where a power switching device is used to energize a short circuit as shown in Fig. 3.33.

As explained in Sect. 3.1.1, the short circuit current can be determined by solving the governing differential equation with consideration of current continuity in the short circuit inductance:

$$\begin{cases} L_{sc}\frac{di_{sc}}{dt} = \sqrt{2}U\cos(\omega t) & t > t_0 \\ i_{sc}(t_0) = 0 \end{cases} \tag{3.58}$$

The short circuit current is then:

$$i_{sc}(t) = \frac{\sqrt{2}U}{L_{sc}\omega}[\sin(\omega t) - \sin(\omega t_0)] \tag{3.59}$$

Under the simplifying assumption that the arcing voltage is constant over the whole period of the pre-strike, the dissipated energy can be calculated using (3.57) and (3.59):

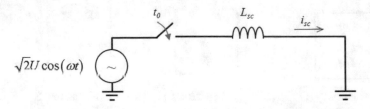

Fig. 3.33 A simple circuit demonstrating the making under short circuit condition

$$E_{dissipation} = \int_{t_0}^{t_{touch}} i_{sc} \cdot u_{arc} dt$$

$$= \frac{\sqrt{2}U}{L_{sc}\omega} \times u_{arc} \left[\frac{1}{\omega} (\cos(\omega t_0) - \cos(\omega t_{touch})) - (t_{touch} - t_0) \sin(\omega t_0) \right]$$

(3.60)

The rate of rise of the short circuit current takes its highest value if the break-down occurs at the peak of the system voltage $t_0 = 0$. Under the assumption that the breakdown occurs at a gap distance of d, and the closing velocity of the contacts is V, the dissipated energy can be expressed as:

$$E_{dissipation} = \frac{\sqrt{2}U}{L_{sc}\omega} \times u_{arc} \left[\frac{\left(1 - \cos\left(\frac{\omega d}{V}\right)\right)}{\omega} \right]$$

(3.61)

For this simplified case, the expression for the total dissipated energy shows a non-linear dependency to the closing velocity of contacts. For a switching component operating in a grid with the frequency of 50 Hz, and assuming that $d = 5$ mm, an increase of the closing velocity from 1 to 5 m/s results in a 96% reduction of the dissipated energy.

Besides the pre-strikes and the effects of the associated energy dissipation, closing operations under short circuit condition also expose the switchgear to very high mechanical stresses as the peak value of the current can by far exceed the amplitude of the AC short circuit current peak value. The electromagnetic (Lorentz) forces are proportional to square of the instantaneous current flowing through the device. As it can be seen from (3.59), the peak value can reach almost twice that of a symmetric AC short circuit current. This results in four times higher mechanical stresses on the switching device compared with the case of making of a symmetric AC short circuit current.

So, depending on the time of breakdown, a short circuit making operation can result in very large mechanical forces (reaching their highest values when $t_0 = \pi/\omega$) and/or high dissipated arcing energies during a pre-strike (reaching their highest values when $t_0 = 0$).

3.2.1 Making Under Short Circuit from a Testing Perspective

According to the IEC standards for circuit breakers, the peak value of the short circuit current during testing should be 2.5 or 2.6 times of the rms value of the rated AC short circuit current, for network frequencies of 50 and 60 Hz, respectively. The time constant of the waveform (representing the decay of the DC component) should be around 45 ms.

One of the situations, where a short circuit current making may occur, is under automatic reclosing operations of circuit breakers. As many faults in transmission and distribution systems are of a temporary or intermittent nature (e.g. lightning strokes, overhead lines touching each other or coming in temporarily contact with trees or other objects), the initial fault cause may be removed by itself after the short circuit current is interrupted. Therefore, in many cases it is possible to energize the system again. If the nature of the fault is of a permanent type, the re-energizing of the system results in short circuit current. To cover this type of operational stresses for the circuit breakers intended to be used with auto-reclosing feature, the switching sequences [e.g. open-close open (O-CO)] are combined in many cases with the short circuit making tests.

The focus of the test procedures employed for verifying short circuit current making ability of circuit breakers is on the high mechanical stresses, which occur during the short circuit making tests with large DC components. The reason is, as explained above, the energy dissipation in pre-strike switching arc by making operation is less of a concern for circuit breakers.

In load break switches and earthing switches, in contrast, the welding problem of contacts by making short circuit currents is more pronounced, so that the maximum number of making operations of a load break switch is in many cases the most important limiting parameter. According to the standards [4], load break switches at different endurance classes have to be able to make the rated short circuit currents for three to five times. The tests cover both the maximum current, i.e. asymmetrical short circuit current with the highest amplitude, which occurs if the pre-strike happens near voltage zero, as well as, the maximum energy case, where the short circuit current is symmetrical and pre-strike happens at the peak of the voltage.

3.2.2 Test Methods

The short circuit current flowing through the power switching device plays a central role in evaluation of its capability to perform closing operation under short circuit current. The initiation of the real pre-strike in the switching object is not possible without application of high voltage to the power switching device prior to its closing operation.

Moreover, to reproduce the real stresses applied to a power switching device during making of a short circuit current, it is necessary to generate both high voltage and high current. This leads to very large amounts of power (as for the case of current interruption tests), if a single power source is used to supply the necessary current and voltage.

Fortunately, high voltage and high current are not necessary at the same time, so that two separate power sources may be used in the same way as the scheme used for testing of current interruption capability of circuit breakers.

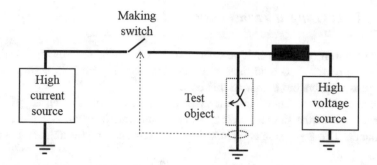

Fig. 3.34 Schematics of a synthetic short circuit current making test circuit

Figure 3.34 shows the basic working principle of a synthetic short circuit making test circuitry. A high voltage source (e.g. a charged capacitor) is directly connected to the terminals of the under test power switching component. A test cycle starts by sending a close command to the test object, when the gap distance between the contacts of the switching device reaches a critical value, a breakdown (pre-strike) is initiated. As the result, current is supplied by the high voltage source (e.g. due to the discharge of the capacitor), this current is detected and a close command is sent to the making switch connecting the high current source and the test object. If the delay between the current flow through the high voltage source and closure of the making switch is small enough, the full short circuit current flows through the previously generated pre-arc channel.

The high current source can be a generator, a grid-connected power source or simply a pre-charged capacitor bank discharged in an inductance. Typical delays of mechanically operating switches are very large. On the other hand, the charge carrying capability of spark gaps is limited. Therefore, the making switch is usually realized as parallel combination of a very fast triggered spark gap and a mechanically operating switch.

3.3   Energization of Loads

One of the basic functions of a switching device is to energize different parts of a network. By closing of a switch, the network configuration abruptly changes and due to energy exchange between different energy storing elements of the power network, transient overvoltage and overcurrent may be produced.

Although the closing operation under load current usually presents no critical stress to the switching device itself, the generated transients may stress other grid components. Such closing switching transients are considered in detail in this section.

3.3.1 Energizing a Transmission Line

A very common operation in a power network is to energize transmission lines by closing one of the switching devices in a substation. Figure 3.35 shows a simple single phase equivalent circuit of this case.

In this circuit, the Thevenin equivalent circuit represents the whole grid on the left hand side of the circuit breaker, and the transmission line is replaced by its capacitance. The governing equations for this circuit can be formulated as follows:

$$
\begin{cases}
L_{sc}C_{line}\frac{d^2 u_{line}}{dt^2} + u_{line} = \sqrt{2}\,U\,\cos(\omega t) \\
u_{line}(t_0) = u_0 & t > t_0 \\
\frac{du_{line}}{dt}(t_0) = 0
\end{cases}
\tag{3.62}
$$

where u_0 is the initial voltage of the transmission line due to the trapped charges at the time of energization. The general solution of the differential equation for the voltage between the transmission line and ground u_{line} contains two components, one with power frequency ω and the other with the frequency $\omega_1 = \frac{1}{\sqrt{L_{sc}C_{line}}}$:

$$
u_{line}(t) = A\,\cos(\omega t) + B\,\sin(\omega t) + A_1\cos(\omega_1 t) + B_1\sin(\omega_1 t) \quad t > t_0 \tag{3.63}
$$

The coefficients of the power frequency terms can be found by replacing the general solution into the differential equation as:

$$
A = \frac{\sqrt{2}U}{1 - L_{sc}C_{line}\omega^2} \quad B = 0 \tag{3.64}
$$

The coefficients A_1 and B_1 are calculated by application of the initial conditions as:

$$
A_1 = u_0\cos(\omega_1 t_0) - \frac{\sqrt{2}U}{1 - L_{sc}C_{line}\omega^2}\left[\cos(\omega_1 t_0)\cos(\omega t_0) + \frac{\omega}{\omega_1}\sin(\omega t_0)\sin(\omega_1 t_0)\right]
$$

$$
B_1 = u_0\sin(\omega_1 t_0) - \frac{\sqrt{2}U}{1 - L_{sc}C_{line}\omega^2}\left[\sin(\omega_1 t_0)\cos(\omega t_0) - \frac{\omega}{\omega_1}\sin(\omega t_0)\cos(\omega_1 t_0)\right]
$$

$$
\tag{3.65}
$$

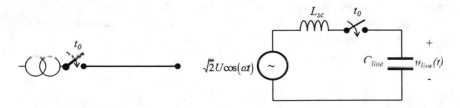

Fig. 3.35 Simplified equivalent circuit for the case of energization of a power line

The coefficients of the terms with the Eigen frequency of the network (A_1 and B_1) are very dependent on the time of energization as well as on the initial voltage of the transmission line caused by trapped charges. In many network configurations, the Eigen frequency of the network is much larger than the power frequency ($\frac{\omega}{\omega_1} \ll 1$). Under the worst case if the closing of the switch occurs at the peak of the network voltage ($t_0 = 0$) and the transmission line has a trapped charge with the opposite polarity, large overvoltage may be produced during the energization operation of a transmission line, see Fig. 3.36. Note that the damping of oscillations is not considered here.

In case of more complex network configurations, e.g. if many transmission lines are connected to a bus bar and energization of one of the transmission lines takes place, the waveform of the transient overvoltage may be much more complicated with more than one oscillation frequency.

The generated overvoltage corresponds to an overcurrent (large charging current) because the high rates of change of the voltage over the capacitance of the transmission lines cause a current flow. For the simple case considered in Fig. 3.35, this charging current can be expressed as:

$$i_{line}(t) = -\frac{\sqrt{2}U\omega}{1 - L_{sc}C_{line}\omega^2}\sin(\omega t) + B_1\omega_1\cos(\omega_1 t) - A_1\omega_1\sin(\omega_1 t) \quad t > t_0$$

$$(3.66)$$

The amplitude of the higher frequency term of the charging current could be much larger than the power frequency charging current. Although this term decays

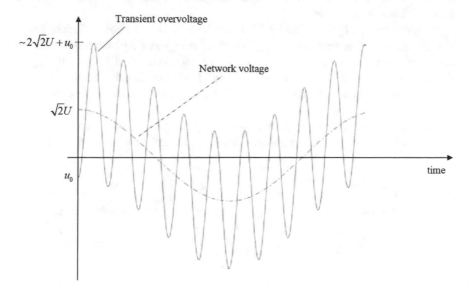

Fig. 3.36 Overvoltage on a transmission line with opposite polarity trapped charge by its energization

after a while due to the resistive losses of the transmission lines and its amplitude is normally much smaller than the short circuit current, it could contribute to high dissipation in the pre-arcing phase, in the same way as previously discussed for the case of making of a short circuit current. As the frequency of this current is much larger than the power frequency, the term $\frac{\omega d}{V}$ in (3.61) becomes even larger than for the case of making currents with the power frequency. This results in larger dissipations during pre-arcing when energizing a transmission line and to a probable contact welding or contact surface deterioration, especially if this operation is performed frequently.

Such overvoltage and high frequency charging current are of importance for switching devices operating both in overhead line and cable based transmission systems. Consequently, in the standards for circuit breakers, cable charging and line charging test duties are included.

3.3.2 Energizing a Capacitor Bank

Capacitor banks are normally used in power networks to compensate the reactive (inductive) load currents and to stabilize the network voltage level within a desired range. For this purpose, it is necessary to connect the capacitor banks, whenever the load is high, and disconnect them from the network when the load is low. Hence, quite often the daily load pattern causes capacitor bank switchgear to operate frequently. In this section, the stresses accompanied with such switching operations are considered with basis in the circuit diagram shown in Fig. 3.37, where a capacitor bank is energized.

By considering Figs. 3.35 and 3.37, it can be seen that the case of energization of a capacitor bank is very similar to the case of energizing a transmission line. All the equations derived in the last section can be applied for this case, if C_{line} is replaced by C_b. Due to the high amplitude, high frequency inrush current associated with energization of a capacitor bank, welding and contact surface deterioration are of concern.

In some networks, there are more than one capacitor bank connected to a busbar. By energization of the first capacitor bank, the configuration is as shown in

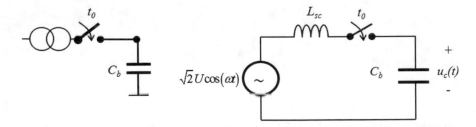

Fig. 3.37 Simplified equivalent circuit for the case of energization of a single capacitor bank

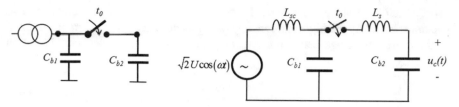

Fig. 3.38 Simplified equivalent circuit for the case of energization of a capacitor bank in back to back configuration

Fig. 3.37, but for the case of connecting the second capacitor bank, the circuit diagram is different, see Fig. 3.38.

The governing equations for this second case can be expressed as follows:

$$\begin{cases} C_{b2}\frac{du_c}{dt} + \frac{1}{L_s}\int (u_c - u_1)dt = 0 \\ C_{b1}\frac{du_1}{dt} + \frac{1}{L_{sc}}\int (u_1 - \sqrt{2}U\cos(\omega t))dt + \frac{1}{L_s}\int (u_1 - u_c)dt = 0 \end{cases} \quad t > t_0 \quad (3.67)$$

where u_1 is the voltage across the first capacitor bank (C_{b1} in Fig. 3.38). By combining these two differential equations, the following differential equation for the voltage across the second capacitor bank (u_c) can be derived:

$$L_s L_{sc} C_{b1} C_{b2}\frac{d^4 u_c}{dt^4} + [(C_{b1} + C_{b2})L_{sc} + C_{b2}L_s]\frac{d^2 u_c}{dt^2} + u_c = \sqrt{2}U\cos(\omega t) \quad (3.68)$$

Under assumption that the voltage of the first capacitor bank has reached its steady state condition before energizing the second capacitor bank, the necessary initial conditions for (3.68) are as follows:

$$u_c(t_0) = 0$$

$$\frac{du_c}{dt}(t_0) = 0$$

$$\frac{d^2 u_c}{dt^2}(t_0) = \frac{\sqrt{2}U\cos(\omega t_0)}{L_s C_{b2}(1 - \omega^2 L_{sc}C_{b1})} \quad (3.69)$$

$$\frac{d^3 u_c}{dt^3}(t_0) = -\frac{C_{b1}}{L_s C_{b2}}\frac{\sqrt{2}U\sin(\omega t_0)}{(1 - \omega^2 L_{sc}C_{b1})}$$

In addition, it has been assumed that the initial voltage of the second capacitor bank is zero at the time of energization. The voltage across the second capacitor bank and the inrush current have terms with three different frequencies, i.e. network frequency ω and two Eigen frequencies of the differential equation ω_1 and ω_2. If the capacitor banks have capacitances values of the same order of magnitude and the short circuit inductance L_{sc} is much larger than the inductance connecting the two capacitor banks L_s, the highest frequency component of the inrush current is related to the energy exchange between capacitor banks through the inductance L_s with a frequency:

$$\omega_1 = \frac{1}{\sqrt{L_s \frac{C_{b1}C_{b2}}{C_{b1}+C_{b2}}}} \tag{3.70}$$

As the inductance between the two capacitor banks is normally much less than the short circuit inductance of the network, the frequency and amplitude of the inrush current are much higher than for the case of energization of a single capacitor bank. Depending on where on the voltage waveform switchgear contacts close, the amplitude of this high frequency inrush current can take very high values.

In the relevant IEC standards, an inrush current with the frequency of 4250 Hz and an amplitude of 20 kA is considered for the case of so-called back-to-back capacitive current switching (Fig. 3.38), while the inrush current of a single capacitor bank testing should have a frequency of a few hundred hertz and an amplitude of a few kilo-amperes.

This high frequency, high amplitude current is very critical from the perspective of the energy dissipated in the pre-arcing phase. Hence, the risk of a contact welding is rather high and the contact surfaces may be severely degraded.

3.3.3   Energizing of a No Load Transformer

Transformers and reactors with magnetic saturable (i.e. iron) cores are used extensively in power networks. Under normal operation of transformers, the magnetic fields of both windings cancel each other almost completely, and the transformer operates in the linear range of the ferromagnetic material. By energization of a transformer or reactor, as there is no counter magnetic field to cancel the magnetic field produced by current flow through its primary winding, the magnetic flux density may increase in such a way that the ferromagnetic core enters the saturation region. This results in drastically decreased inductance and causes, in turn, an additional current increase. In the following single phase example, this phenomenon is treated in detail. Consider a transformer connected to a power network with a short circuit inductance of L_{sc} as shown in Fig. 3.39.

The reactor or the no-load transformer is simply replaced by its magnetization inductance. As the short circuit impedance is normally much less than the magnetization impedance, the voltage drop over the short circuit impedance may be neglected, and it can be assumed that the entire network voltage is applied over the inductor terminals by closing the switch.

Current and voltage of the non-linear magnetization inductor can be linked together using the following equations:

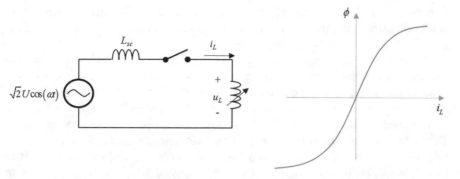

Fig. 3.39 Simple circuit for explanation of inrush current by energization of no load transformer and reactors with saturable ferromagnetic core

$$i_L = f(\phi)$$

$$\phi = \phi_0 + \int\limits_{t_0}^{t} u_L(t)dt \tag{3.71}$$

Here ϕ is the magnetic flux through the ferromagnetic core of the transformer or the inductor, ϕ_0 is the residual flux of the core, i_L and u_L are the magnetization current and voltage across the inductor, respectively.

If, for example, $\phi_0 = 0$ (no residual magnetic flux in core at the time of energization), then the magnetic flux of the core (ϕ) can reach twice of its maximum steady state value, if the inductor is energized at voltage zero. Considering the very nonlinear relationship between i_L and ϕ as shown schematically in Fig. 3.39, this can result in a significant increase of the magnetization current at the time of energization. On the other hand, energization at the voltage peak results in a maximum magnetic flux of its steady state value (the working point stays in the non-saturated part of the magnetization curve) and consequently results in no overcurrent by energization. This very simple example shows the importance of where on the supply voltage waveform the energization takes place.

High inrush currents stress the transformer or reactor mechanically, so the repetition of such stresses may in the long term damage the transformer.

3.3.4 Synchronized Closing

As shown above, during energization of energy storing elements like capacitors, transmission lines as well as no-load transformers or reactors, transient overvoltage and/or overcurrent can be generated depending on the time of energization. In conventional power switching components, the time of contact opening or closing is a random variable and therefore, it is a certain probability that the worst case scenarios in case of energization of energy storing elements occur, resulting in the most undesirable overvoltage and overcurrent.

To limit the overvoltage and overcurrent transients, one possibility is to control the mechanical closing time of the circuit breaker contacts. For this purpose, the state of the energy storage element, e.g. trapped charge of the transmission line, initial charge of a capacitor bank and the remanence magnetic field of a transformer core, as well as the voltage on the source side have to be taken into consideration. Depending on the state of the energy storage element, the best time to energize it, with consideration of minimizing the transient overvoltage and overcurrent, has to be calculated. By taking into account the time delay of the operating mechanism of the switching device, the optimum time for sending a closing command can be determined.

In practice, as the electrical closing happens by the breakdown of the gap between two contacts of the switching device, the dielectric strength behaviour of the switching gap under a closing operation becomes important. This is described by the so-called rate of decrease of dielectric strength (RDDS), which is very dependent on the travel curve of the moving contact, on the contact geometry as well as on the insulating medium. Another practical concern in the context of controlled closing is the scatter of the delay time, known as jitter, due to change of environmental and/or operational conditions. This implies that the delay time between sending the closing command to the switching device and the electrical closure of the circuit breaker is not a constant parameter. The jitter is dependent on the type of driving mechanisms. For example, the operating time of spring drives is far less sensitive to temperature variations than hydraulic drives. Furthermore, the jitter can be reduced by choosing better controllable drive mechanisms like permanent magnet or stepped motor drives. The random scatter of the closing delay time normally becomes lower if the closing velocity is increased.

Obviously, the controlled closing feature of a switchgear normally adds cost and complexity to the drive mechanism. In addition, a control unit with some measurement devices is required, which makes it possible to find the optimum time of energization. Such an added feature causes severe overvoltage or overcurrent transients to be avoided, and this improves the long-term reliability of the switchgear and nearby components.

3.4   Interruption of Load Currents

Energization and de-energization of different types of loads in a power network occur very often. Therefore, it is quite important for power switching devices to be designed in such a way that they perform this type of switching operations reliably.

3.4.1 Interruption of Inductive–Resistive Currents

For some devices (load break switches), interruption of inductive–resistive load currents presents the highest interruptible current, whereas for some other (circuit breakers), this is a quite easy switching duty. In many applications, the nature of loads is typically inductive and resistive, but for some special application, switching of capacitive loads is also of importance. The latter is covered in the next section. In this section, the inductive–resistive load switching is considered. Figure 3.40 can be used to explain the stresses applied to the load break switch.

Unlike the terminal short circuit current interruption, by interruption of load currents, transient voltages appear at both sides of the switch, see Fig. 3.40. The applied TRV then becomes the difference between these two transient voltages.

The load impedance Z_l and the source impedance Z_s are comparable in size and therefore even with the power switch in closed position, the entire source voltage is not applied to the load. The relevant IEC standard [4] sets the amplitude and type of the source and load impedances to be used for type testing. The load impedance has to be a parallel combination of inductive and resistive elements with a load factor of about 0.7, while the source impedance is a series combination of inductive and resistive elements having an impedance of 15% of the total network impedance and with a load factor of 0.15. The parallel branch consisting of a resistor and a capacitor is used to generate the source side TRVs. If the current flowing through the parallel impedance of Z_p under power frequency voltages is neglected, the following expressions can be derived:

$$Z_l + Z_s = Z_{total}$$
$$Z_s = 0.15|Z_{total}|\angle 78.46°$$
$$Z_l = |Z_l|\angle 45.6°$$

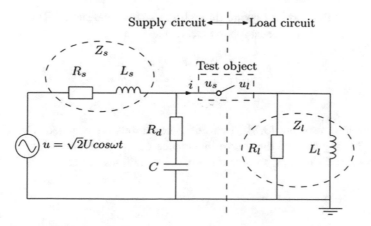

Fig. 3.40 Single-phase circuit for mainly active load switching test [5]

If we assume $|Z_l| = k|Z_{total}|$ and combine it with above equations, then k is calculated to be approximately 0.87. This means independent of the voltage or current level, the load impedance has to be about 87% of the total network impedance. In the same way, the power factor of the total network impedance can easily be calculated to be approximately 0.64.

The load side transient recovery voltage is an overdamped voltage waveform formed by decaying of the initial current of the load inductance through the load resistance. After current zero crossing, the governing equation for the load side circuit can be written as:

$$\frac{1}{L_l} \int u_l dt + \frac{u_l}{R_l} = 0$$
$$u_l\left(t_{cz}^+\right) = 0.64 \times \sqrt{2}\,U \tag{3.72}$$

By solving this first order differential equation, load side transient recovery voltage is derived:

$$u_l(t) = 0.64 \times \sqrt{2}\,U\left[1 - \exp\left(-\frac{R_l}{L_l}(t - t_{cz})\right)\right] t \geq t_{cz} \tag{3.73}$$

For the conditions of the so-called *mainly active load switching duty* of the IEC standard, the time constant of the decaying voltage is about 3 ms and independent on the current and voltage levels.

The source side transient recovery voltage can be calculated by considering the governing equation of the source side circuit after current zero. The voltage has a phase shift of about 50.2° to the current and therefore at the current zero crossing, the voltage can be expressed as $\sqrt{2}\,U\,\cos(\omega t + 50.2°)$:

$$L_s \frac{di_s}{dt} + (R_s + R_d)i_s + \frac{1}{C} \int i_s dt = \sqrt{2}\,U\,\cos[\omega\,(t - t_{cz}) + 50.2°] \quad t > t_{cz} \tag{3.74}$$

Here, i_s is the current flowing through the source impedance. The source side transient recovery voltage can then be calculated as:

$$u_s(t) = R_d i_s + \frac{1}{C} \int_{t_{cz}}^{t} i_s dt \tag{3.75}$$

To be able to solve (3.74), two initial conditions are required. Continuity of current flowing through the source inductance L_s and continuity of voltage across the capacitor C, are used to derive the initial conditions at t_{cz}^+:

$$i_s\left(t_{cz}^+\right) = 0$$

$$\frac{di_s}{dt}\left(t_{cz}^+\right) = \frac{u\left(t_{cz}^+\right) - u_C\left(t_{cz}^+\right) - (R_d + R_s)i_s\left(t_{cz}^+\right)}{L_s}$$

$$= \frac{\sqrt{2}\,U\,\cos(50.2°) - 0.87 \times \sqrt{2}\,U\,\cos(45.6°)}{L_s} \approx 0.031 \times \frac{\sqrt{2}\,U}{L_s}$$

$$(3.76)$$

The voltage at the capacitor just before current interruption is equal to $0.87 \times \sqrt{2}\,U \cos[\omega(t - t_{cz}) + 45.6°]$, if the current flowing through the RC parallel branch is neglected.

A typical waveform of the TRV, i.e. the difference between u_s and u_l, is shown in Fig. 3.41.

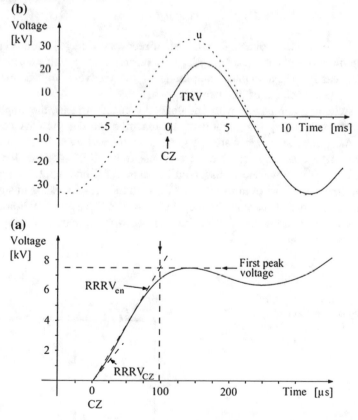

Fig. 3.41 Typical transient recovery voltage in the case of mainly active load current interruption **a** the complete picture **b** the first transient just after current zero [5]

3.4.2 Interruption of Capacitive Currents

This switching duty occurs when capacitor banks, no-load lines and cables are disconnected. The currents are relatively small, typically between some tens and a few hundred amperes. The circuit diagram for interruption of a capacitive load current is shown in Fig. 3.42.

In this simple equivalent circuit, C_1 and C_2 represent the stray capacitance of the source side network and the load capacitance, respectively. The current amplitude is mainly determined by u_p and C_2. C_1 is assumed to carry a negligible current prior to the interruption. Figure 3.44 shows waveforms on both sides of the breaker on a certain case of capacitive load current interruption. The voltage in a capacitive circuit is at its maximum at the time of current zero crossing. If the current is interrupted at this point, the voltage on the load side stays high due to charging of the capacitor; u_r equals the peak value of the power frequency voltage u_p. The voltage on the left side, u_l, follows the network voltage. The recovery voltage applied to the terminals of the switching device is then calculated as:

$$u_b(t) = u_r - u_l = u_p(1 - \cos \omega t) \tag{3.77}$$

where u_p is the network voltage. Rate of rise of recovery voltage is in this case near zero at the time of the current interruption. Therefore, the capacitive current can be interrupted easily. The maximum amplitude of the recovery voltage may reach twice of the peak voltage of the network.

For a switching component with undamaged contact surfaces, the amplitude of the recovery voltage by interruption of a capacitive current presents no critical stresses. All power switchgear are designed to withstand even larger power frequency voltages than this across open contacts (typically 2–4 times the system voltage for 1 min). However, as discussed in Sects. 3.3.1 and 3.3.2, by energizing capacitive loads, the flow of inrush current through the pre-arc results in significant degradation of the contact surfaces (see Fig. 3.43), so that the amplitude of the recovery voltage in the range of twice the rated network voltage peak may in such cases become critical.

Fig. 3.42 Circuit diagram for interruption of a capacitive load; $C_2 \gg C_1$

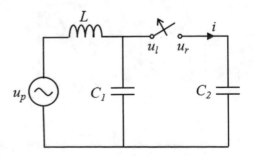

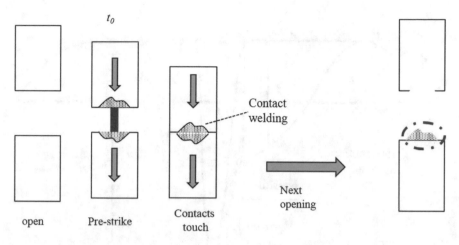

Fig. 3.43 Degradation of contact surface of a power switching component by energizing and de-energizing of capacitive loads

If a re-strike occurs as u_b reaches its maximum, the load side voltage starts rapidly oscillating around the source voltage. If the small discharge current is interrupted at its first current zero crossing, the charging voltage of the capacitor C_2 reaches almost three times of the peak rated voltage of the network. The amplitude of the voltage across the contacts can become about four times of the system voltage, see Fig. 3.44. On the other hand, a breakdown within the first quarter of a cycle after current interruption, does not lead to larger charging voltages of the capacitor. Therefore, there is a differentiation between breakdowns within the first quarter of a cycle after current interruption, which are called re-ignitions, and breakdown after that time, which are called re-strikes.

Even larger voltages may build up if more interruptions and re-strikes occur. These over-voltages may lead to a breakdown between the phases and/or to ground in equipment nearby the switchgear. To avoid such problems it is crucial that capacitive currents are interrupted using switchgears that do not re-strike.

Hence, it is clear that the interruption of capacitive current itself is not the major problem, but the main problem is due to the degradation of the contacts by energization of the capacitor loads, which may lead to a re-strike by the application of the transient recovery voltage. This explains why capacitive current switching with the same current amplitude is more critical if the capacitor bank is used in a back-to-back configuration.

Re-strike free switching of capacitive currents represents a major challenge for many circuit breakers and having this feature is an indication of their superior design and performance.

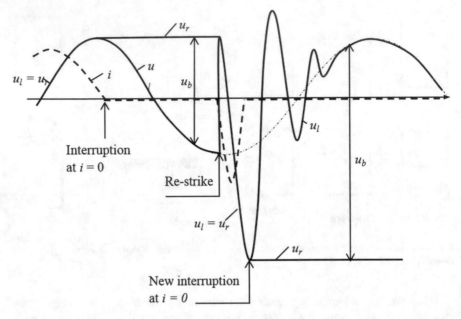

Fig. 3.44 Voltage escalation during interruption, re-ignition and a new interruption in a capacitive load circuit

3.4.2.1 Capacitive Current Switching from Standards and Testing Point of View

As discussed above, there are two main issues associated with capacitive current switching. These are degradation of the dielectric strength of the switching device due to the flow of inrush current through pre-arc by energization of the capacitive loads, and the recovery voltage in the form of $1 - \cos\omega t$ applied to the terminals of the switching device by current interruption.

The IEC standards describe the conditions for testing of the capacitive current interruption capability of switchgear. These can be categorized in three different applications:

- Line charging current breaking tests
- Cable charging current breaking tests
- Capacitor banks switching tests.

Line charging is usually associated with high voltage switching devices (i.e. for rated voltages above 72.5 kV), where the connected transmission systems are in many cases overhead transmission lines. For medium voltage switching devices (i.e. for rated voltages below 72.5 kV), the dominating transmission systems is, however, cable and therefore the cable charging test duty is to be considered.

Capability to switch capacitor banks, i.e. single or back-to-back, is normally an added feature, so that if a switching device is claimed to be suitable for switching of capacitor banks, it has to perform many closing and opening operations without any re-strikes. According to the IEC standard, there are two different categories of capacitive current switching capability, namely class C1 (low re-strike probability) and class C2 (very low re-strike probability), where the number of re-strikes during the whole test is decisive.

Although application of transient recovery voltage after interruption of the capacitive load current is the same for all test duties, major differentiations are the frequency and amplitude of the inrush current by closing operation. The recommended frequency and amplitude of the inrush current in case of back-to-back capacitor switching tests are 4250 Hz and 20 kA. For the other test duties, there is no explicit recommendation in the standard [3].

3.4.2.2   Test Methods

According to the standards, both three phase and single phase tests, may be performed to evaluate the capability of the switching devices to cope with the operational stresses related to the switching of capacitive loads.

Very high currents and high voltages are needed for capacitive current switching test. However, as these stresses are not applied at the same time, it is possible to provide large inrush currents and large voltages from two different power sources. In the IEEE guideline for synthetic testing of capacitive current switching [6], different methods have been proposed. A simple test circuit shown in Fig. 3.45, demonstrates a circuit with one AC source and a tuned circuit current branch.

Each test cycle has two parts; the first part begins with closing of the test object by flow of inrush current through the pre-arc. One simple circuit to generate the inrush current is to discharge a capacitor through a low loss inductor as shown in Fig. 3.45. During this part of the test, the high voltage parts are disconnected. As explained earlier in Sect. 3.1.4.2, as the high current part of the synthetic short circuit current interruption test circuit, the capacitance, inductance and charging

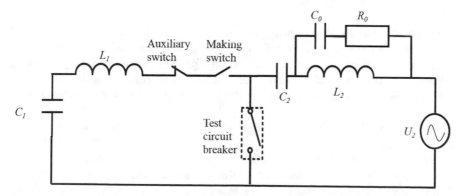

Fig. 3.45 A synthetic test circuit for capacitive current switching tests

voltage of the capacitors can control the current amplitude and frequency. The charging voltage of the capacitors should be near the rated voltage of the breaker, otherwise it is not possible to realize the real operational stresses during the pre-arcing.

After closing the test object, it has to be opened while the rated capacitive current is flowing. For this purpose, the high current circuit used to generate the inrush current is disconnected. As stated earlier, the capacitive current is interrupted easily at its zero crossing, because of the very low rate of rise of recovery voltage. In some laboratory setups, the current flowing through the switch during the opening operation is almost zero. This testing condition is claimed to be even tougher than the standard requirements, as there is no smoothening effect of contact surface during arcing. Having a charged capacitor in series with the switching device being tested, ensures that the applied transient recovery voltage has the desired $1 - \cos\omega t$ waveform.

3.4.3 Interruption of Small Inductive Currents

Circuit breakers are mainly designed for interrupting short circuit currents, i.e. large inductive currents. In some cases however, certain phenomena leading to large over-voltages, and possibly also faults, occur when interrupting inductive currents that are *small* compared to the capability of the breaker. Three such cases are studied here: (i) current chopping, (ii) multiple re-ignitions in interrupters with fast dielectric strength recovery, and (iii) virtual current chopping.

These phenomena may typically occur when disconnecting no-load transformers, reactors and rotating machines, especially during start-up. The current may vary from a few tens of amperes in transformers to a few thousand amperes during start-up of large motors.

3.4.3.1 Current Chopping

A high current interruption capacity is in many cases achieved by making sure that the arc is efficiently cooled by a flow of cold gas. In some circuit breaker technologies, for example oil breakers and certain types of SF_6 breakers, the gas flow is generated by the arc itself. The switchgear may then be constructed in such a way that the gas blast is stronger when interrupting large currents than with smaller currents.

In other breaker technologies, e.g. air blast circuit breakers and older SF_6 switchgear the gas flow is largely independent of the magnitude of the current being interrupted; the flow intensity is the same regardless of the switching duty.

Consequently, in some cases the cooling effect can become so powerful that during interruption of small currents the electric arc does not remain stable until the natural current zero crossing. The arc becomes unstable and may quench before current zero. This for the most occurs in switchgear where the gas flow is independent of the current magnitude. The phenomenon is called *forced interruption* or *current chopping*, see Fig. 3.46.

The term "current chopping" is used because the transition from conducting to insulating state proceeds so fast, compared to the system frequency, that it can be considered a step in current. The current amplitude at the moment of chopping varies from one type of switchgear to the next, but is generally between one and a few tens of amperes.

Only interruption of small currents can lead to current chopping. Higher currents will cause the arc to remain much more stable until the point of natural current zero. Chopping does not lead to problems when interrupting small capacitive currents. When interrupting small inductive currents, on the other hand, problems may arise from the chopping. These are mainly due to generation of large over-voltages in the system, as will be explained in the following sections.

Current and voltage during current chopping is studied with basis in the circuit diagram in Fig. 3.47.

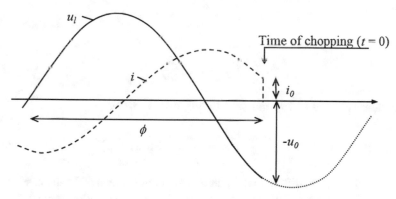

Fig. 3.46 Current and voltage during interruption of a small inductive current where current chopping occurs. The current is chopped when its magnitude is i_0 and the associated voltage is $-u_0$

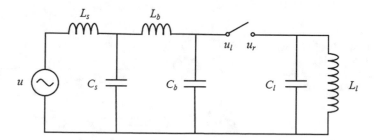

Fig. 3.47 Circuit diagram when interrupting small inductive currents

L_s and C_s are the short circuit inductance and source side system capacitances, respectively, whereas L_b and C_b represent the inductance and capacitance on the busbar side of the switchgear. The inductive load in the circuit (transformer, reactor, motor etc.) is L_l, and the load side capacitance is C_l.

The source side system capacitance C_s, is far greater than the busbar capacitance C_b. In the cases of fast voltage transients (high frequencies), C_s essentially acts as a short circuit compared to C_b. The Thevenin impedance of the source side system ($L_s \parallel C_s$) is thus very small, and when considering only fast transients the source side system can, with a reasonable accuracy, be represented by only a voltage source. The circuit diagram is then simplified to that shown in Fig. 3.48.

The recovery voltage across the breaker contacts due to current chopping will now be examined. The voltages on either side of the breaker are calculated separately. It is assumed that the current is chopped prior to current zero crossing at the magnitude $i = i_0$, and that the voltage at this time is $u = -u_0$, see Fig. 3.46.

Based on the diagram of Fig. 3.48, the circuit equations for the busbar side of the contacts after interruption become

$$u - L_b \frac{di_{C_b}}{dt} - u_l = 0 \tag{3.78}$$

and

$$i_{C_b} = C_b \frac{du_l}{dt} \tag{3.79}$$

By differentiating (3.79) and substituting it into (3.78):

$$L_b C_b \frac{d^2 u_l}{dt^2} + u_l = \sqrt{2}\, U \, \cos(\omega t + \phi) \tag{3.80}$$

The circuit and the differential equations are almost identical to the case studied Sect. 3.1.2.2. The only difference is that the time of interruption ($t = 0$) does not come at the natural current zero crossing, and the phase angle ϕ must therefore be included in the system voltage expression (u). The general solution is:

Fig. 3.48 Simplified circuit diagram valid for high frequencies. The source side is represented by a voltage source

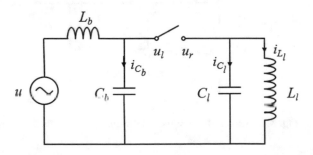

$$u_l = A \sin\frac{t}{\sqrt{L_b C_b}} + B \cos\frac{t}{\sqrt{L_b C_b}} + \sqrt{2}\,U \cos(\omega t + \phi) \qquad (3.81)$$

when it is assumed that $\omega^2 L_b C_b \ll 1$. The initial conditions are

$$u_l(0) = -u_0 \qquad (3.82)$$

and

$$i_{C_b}(0) = C_b \frac{du_l}{dt} = i_0 \qquad (3.83)$$

since the current is forced through C_b at $t = 0$ instead of passing through the breaker. By inserting this into (3.81), the expression for the source side voltage is found:

$$u_l = \left[\sqrt{\frac{L_b}{C_b}} i_0 + \sqrt{2 L_b C_b}\,U\omega \sin\phi \right] \sin\frac{t}{\sqrt{L_b C_b}}$$
$$- \left[u_0 + \sqrt{2}\,U \cos\phi \right] \cos\frac{t}{\sqrt{L_b C_b}} + \sqrt{2}\,U \cos(\omega t + \phi) \qquad (3.84)$$

The last and stationary term of the expression is equal to the source voltage u, and shows, as could be expected, that the voltage on the left of the breaker u_{ls} oscillates towards the source voltage. The frequency f_t of the transient part (the first two terms) is:

$$f_t = \frac{1}{2\pi\sqrt{L_b C_b}} \qquad (3.85)$$

The busbar inductance is normally relatively small, which causes the oscillation to have a high frequency, typically around 10^5 Hz. The amplitude is predominantly determined by the term $i_0 \sqrt{C_b/L_b}$. In most cases, the transient part amplitude is less than the source voltage amplitude, and it will normally be strongly damped.

The voltage waveform on the busbar side of the breaker after current chopping is thus a high frequency transient of moderate amplitude and large damping, super-imposed on the source voltage.

The current is chopped prior to the natural current zero crossing which causes some to be "left behind" or "trapped" on the right side of the breaker (the load side) after the interruption. The energy associated with this current oscillates between being stored in the magnetic field of the inductance L_l and the electrical field of the capacitance C_l.

The peak value of the voltage to the right of the breaker can thus be found by energy considerations. At the time of the interruption the amount of energy stored on the right side is given by:

$$W = \frac{1}{2}L_l i_0^2 + \frac{1}{2}C_l u_0^2 \tag{3.86}$$

where i_0 and u_0 are the current and voltage, respectively, at the time of the current chopping, see Fig. 3.46. The voltage reaches its maximum, U_{max}, when all the energy is stored in the capacitance and can be derived from the expression:

$$\frac{1}{2}C_l U_{max}^2 = \frac{1}{2}L_l i_0^2 + \frac{1}{2}C_l u_0^2 \tag{3.87}$$

giving

$$U_{max} = \sqrt{u_0^2 + \frac{L_l}{C_l} i_0^2} \tag{3.88}$$

When the load is inductive, the L_l/C_l ratio is large and the maximum voltage on the load side after interruption is considerably higher than the system voltage.

By applying standard circuit analysis, an analytic expression of the voltage waveform on the right of the breaker can be determined. As there is no current input from the busbar side after the interruption, the sum of the currents in the two parallel load side branches (see Fig. 3.48) is zero

$$i_{L_l} + i_{C_l} = 0 \tag{3.89}$$

Moreover,

$$i_{C_l} = C_l \frac{du_r}{dt} \tag{3.90}$$

and

$$u_r = L_l \frac{di_{L_l}}{dt} \tag{3.91}$$

By inserting (3.90) and (3.91) into (3.89) and then differentiating, the differential equation for u_r is found:

$$\frac{d^2 u_r}{dt^2} + \frac{1}{L_l C_l} u_r = 0 \tag{3.92}$$

The general solution is on the form:

$$u_r = A \sin\frac{t}{\sqrt{L_l C_l}} + B \cos\frac{t}{\sqrt{L_l C_l}} \tag{3.93}$$

The initial conditions are:

$$u_r(0) = -u_0 \tag{3.94}$$

and

$$i_{L_l}(0) = -i_{C_l}(0) = -C_l \frac{du_r}{dt} = i_0 \tag{3.95}$$

when it is assumed that the capacitive current at the time of the interruption is negligible, i.e., the entire current goes through the load L_l.

The solution to (3.92) thus becomes:

$$u_r = -i_0 \sqrt{\frac{L_l}{C_l}} \sin \frac{t}{\sqrt{L_l C_l}} - u_0 \cos \frac{t}{\sqrt{L_l C_l}} \tag{3.96}$$

Compared to the busbar side the frequency on the load side is, due to the large inductance of the load L_l, far lower, typically around 10^3 Hz. The amplitude on the load side can be considerably greater than on the busbar side. The current and voltage on both sides of the contacts are shown schematically in Fig. 3.49.

The contact gap is exposed to the difference $u_b = u_l - u_r$. This voltage may reach such a large magnitude that the gap re-ignites. Shortly after the re-ignition the current has another zero crossing. The switchgear may then interrupt again, again causing a substantial voltage across the contacts which in turn may give rise to a

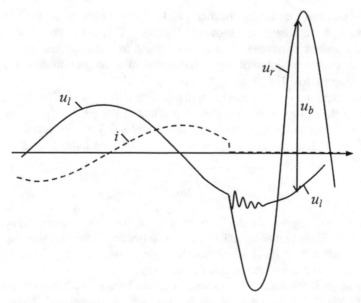

Fig. 3.49 Current and voltage during interruption of a small inductive current when the current is chopped prior to current zero crossing. The transients are shown with considerably lower frequencies than what is generally found in practice

new re-ignition. This may continue, resulting in a series of voltage transients with their amplitudes determined by the dielectric strength of the contact gap. The dielectric strength increases as the contacts move further apart, hence the over-voltages may increase for each re-ignition. A certain amount of energy is dissipated in the circuit at each re-ignition. If it is assumed that this happens so fast that only insignificant amounts of energy is added from the system side, the energy stored in the load side inductances and capacitances decreases until no further re-ignitions occur, until voltage surge arrestors are activated, or until a dielectric failure occurs.

3.4.3.2   Multiple Re-ignitions in Switchgears Having a Rapid Dielectric Recovery

When interrupting small inductive currents over-voltages may also occur when the interruption happens at the natural current zero crossing (i.e. no current chopping). This is studied with basis in the equivalent circuit in Fig. 3.48. The following is assumed:

- The transients are so fast that the voltage to the left of the switchgear, u_l, is considered constant.
- The load side capacitance is far greater than the busbar capacitance; $C_l \gg C_b$.
- The contacts separate just prior to current zero crossing, and the dielectric strength across the gap increases rapidly and linearly with time.

Possible waveforms for the busbar and load side voltages, as well as for the current through the breaker are shown in Fig. 3.50. This development presupposes several re-ignitions across the contacts. The various inductances and capacitances in the circuit interact in different ways before and after re-ignition and resulting in different frequencies. This is illustrated in Fig. 3.51.

The interruption occurs at current zero crossing at the time t_1. The voltage on the source side (left side) of the switchgear remains at the peak value of the source voltage while the voltage on the load side (right side) of the breaker oscillates around zero. Its frequency is determined by the load side impedances:

$$f_l = \frac{1}{2\pi\sqrt{L_l C_l}} \tag{3.97}$$

At t_2, the voltage across the breaker is twice the source voltage u, and it is assumed that this is just sufficient to cause a re-ignition. The load side capacitance C_l is far greater than the busbar (source) side capacitance C_b. Therefore, u_l is dragged down to the voltage across C_l, i.e. to u_r.

The voltage at the breaker, u_l and u_r, then oscillates towards the source voltage u. As the left and right sides of the breaker now are connected, the oscillating frequency is determined by all four impedances in parallel:

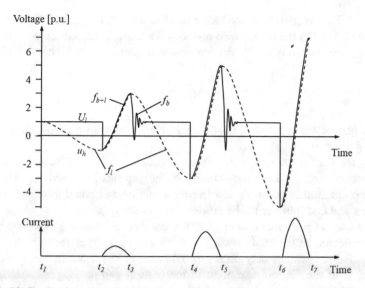

Fig. 3.50 Idealised voltage waveforms during multiple re-ignitions under interruption of a small inductive current. The voltages escalate as the dielectric strength in the gap increases with increasing gap distance. The voltage on the busbar (left side of the breaker) is drawn with *solid lines* and the load side is *dashed*

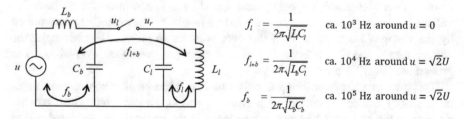

Fig. 3.51 The circuit diagram giving the voltages shown in Fig. 3.50. The inductances and capacitances, which form the basis for the different oscillatory circuits and their typical natural frequencies are shown to the right

$$f_{b+l} = \frac{1}{2\pi\sqrt{\frac{L_b L_l}{L_b + L_l}(C_b + C_l)}} \qquad (3.98)$$

As $L_l \gg L_b$ and $C_l \gg C_b$ the expression is simplified to:

$$f_{b+l} \approx \frac{1}{2\pi\sqrt{L_b C_l}} \qquad (3.99)$$

At t_3, $du/dt = 0$, and the current in this inductive circuit has a zero crossing. The arc extinguishes, the current is interrupted and both the voltages u_l and u_r will start oscillating. The frequency on the load side is determined by L_l and C_l, and is equal to f_l, see (3.97). On the busbar side u_l oscillates towards the source voltage, and as L_b and C_b are relatively small, the frequency is considerably higher on this side.

$$f_b = \frac{1}{2\pi\sqrt{L_bC_b}} \tag{3.100}$$

This part of the waveform is shown in Fig. 3.50 with consideration of some damping.

At t_4, the voltage across the switchgear is four times the peak source voltage and it is assumed that the dielectric strength in the gap has increased just enough for another re-ignition to occur. A new transient oscillating around u and of frequency f_{l+b} starts and continues until the current zero crossing at t_5.

If the contact movement and separation continue so that the dielectric strength continues to increase linearly by 2 p.u every $(1/f_l + 1/f_{l+b})$ seconds, more interruptions and re-ignitions may occur, as indicated in Fig. 3.50. In theory this could continue until the dielectric strength in the contact gap no longer increases, until surge arrestors are activated, or until a fault (spark-over or disruptive discharge) in the switchgear or in other components nearby.

The frequencies f_l, f_{l+b}, and f_b are typically of the order 10^3, 10^4 and 10^5 Hz, respectively. This means that the time periods $t_3 - t_2$, $t_5 - t_4$ etc. are in microsecond range. The voltage distribution in the transformer windings, the reactor or the motor which is being represented by L_l is nonlinear due to these fast voltage transients. The stresses on the winding insulation are essentially determined by the voltage "leap" rather than the over-voltages to ground. For example, the voltage leap in the time interval $t_4 - t_5$ is about 8 p.u., while the over-voltages to ground are "only" 5 p.u.

This kind of voltage build up is only possible if the breaker contacts are separated shortly before the current zero crossing and at the same time the dielectric strength in the gap must be rapidly increasing. If the contacts are separated earlier compared to the current zero crossing, no re-ignition occurs and therefore, no over-voltages either.

If the current to be interrupted is above a certain amplitude the arc will not be extinguished just after contact separation, but most likely at the next current zero crossing. Re-ignitions are thus avoided as the dielectric strength in the gap by then has become sufficiently large, and no over-voltages of the types shown in Fig. 3.50 will build up. If, on the other hand, the current is very small, the over-voltages are limited by the damping of the oscillations on the load side. In the literature, currents in the range 20–600 A are reported to be critical.

Furthermore, it turns out that if the load side oscillating frequency f_l is as low as a few hundred hertz, the dielectric strength increases too fast for re-ignitions to occur. If, on the other hand, f_l is high, the switchgear will be unable to interrupt the high-frequency current at its zero crossings (at t_3, t_5 etc.), and the interruption occurs one half power cycle later, when the contact gap is too great for a re-ignition to occur.

Consequently, the waveforms shown in Fig. 3.50 are limited to a certain current range in combination with a natural oscillation frequency range for the load side oscillations. Furthermore, the descriptions are also somewhat simplified. The situation is somewhat different in a three-phase system; current chopping may influence the waveforms, the dielectric strength may increase non-linearly and other effects may also influence the development.

Still however, such over-voltages when interrupting small inductive currents may occur in different types of switchgear. In particular, they have been associated with vacuum interrupters, which have a very fast build-up of dielectric strength. Protective measures may be necessary, particularly for switchgears connecting small high voltage motors. Surge arrestors with a low protection level are often found to be a suitable remedy.

3.4.3.3 Virtual Current Chopping

In the last section, the multiple re-ignition phenomenon has been examined for a single-phase system. In real applications, the situation is even more complex, as transmission and distribution systems are three-phase systems and there are some couplings between the phases. By occurrence of multiple re-ignitions in one phase, the high frequency currents described in Sect. 3.4.3.2, may flow through the other two healthy phases due to galvanic coupling [7] (e.g. through the neutral point) and capacitive coupling (e.g. through the earth capacitances). This is shown in Fig. 3.52.

If the amplitude of the high frequency currents initiated by re-ignition in one phase is comparable to the power frequency current of the other phases, it may be possible to have very high frequency current zero crossings in healthy phases. This situation is shown schematically in Fig. 3.53. In this case, high frequency currents due to re-ignition in phase S, are coupled to the healthy phases R and T. The

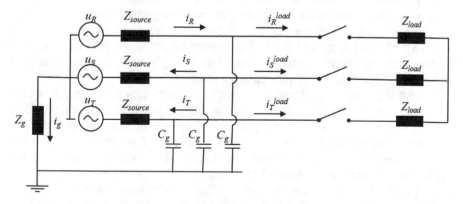

Fig. 3.52 Coupling between three phases due to galvanic and capacitive current paths

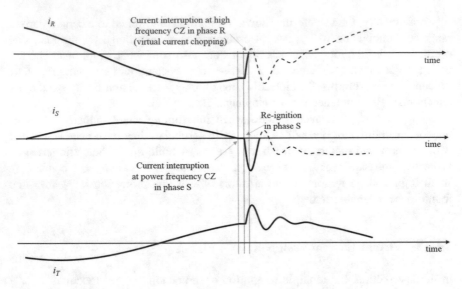

Fig. 3.53 Current waveforms in all three phases by occurrence of multiple re-ignitions in one phase

amplitude of the high frequency current flowing through the healthy phases is so large that a high frequency current zero crossing may occur.

Some switching technologies (e.g. vacuum circuit breakers) can interrupt the very high frequency current at its current zero. By interruption of current with very large $\frac{di}{dt}$, a large voltage is generated over the system inductances as described in Sect. 3.4.3.1. From this perspective, it can be seen that in three-phase networks, current chopping like phenomena may occur, when one phase shows multiple re-ignition behaviour. This interaction and combined way of generation of over-voltages is called *virtual current chopping*.

If the switching device is not able to interrupt the high frequency currents at their zero crossings, the current continues to flow and is interrupted at the power frequency current zero crossing. In this case, no overvoltage due to the virtual current chopping is generated. From this perspective, very superior current interruption capability of a switching device, which is a desired characteristic when interrupting high currents and dealing with steep recovery voltages, can be undesirable as it under certain conditions may generate overvoltages.

The amplitude of overvoltages generated in this way, may be much higher than those generated by current chopping and multiple re-ignitions in a single phase system. In practice, overvoltages larger than 10 per unit have been observed.

Not only the amplitude of overvoltages are very high, but their rise times or fall times are very short. In some power components like as motors and generators, application of very steep overvoltages may introduce additional problems, as capacitances to ground cause an uneven voltage distribution along the winding. A major part of the overvoltage is applied to the first section of the winding.

Therefore, even a rather low amplitude high frequency overvoltage may lead to overstressing the insulation system in motor and transformers potentially damaging these components.

This is the reason why many switchgear manufacturers recommend using appropriate overvoltage protecting devices when vacuum circuit breakers are employed to switch motors and no-load transformers.

Exercises

Problem 1

The circuit of Fig. 3.54 shows a three-phase system with a three-phase-to-ground terminal short circuit.. The breaker has received a command signal for opening; the arc in phase R has extinguished, and this pole can be considered open. The two other circuit breaker poles of phases S and T continue to carry current through an electric arc.

The network neutral point is connected to ground through the impedance L_g, and the short circuit impedance is L. The capacitance between phases is C_p, and the phase-to-ground capacitance is C_g. The source voltage of phase R is given as: $u_{pR}(t) = \sqrt{2} \cdot U_p \cdot \cos(\omega t)$.

    1.1 The single-phase equivalent circuit seen from the *first pole to clear* (between R and R') is shown in Fig. 3.55.

          Show that the equivalent inductance and capacitance become:

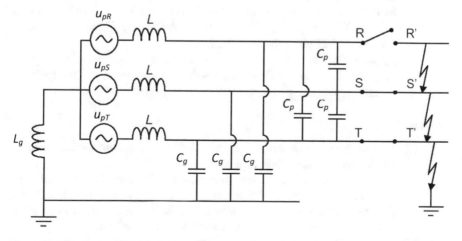

Fig. 3.54 First figure of problem 1

Fig. 3.55 Second figure of
problem 1

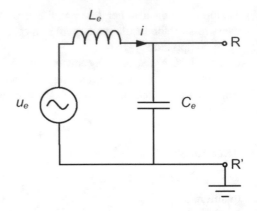

$$L_e = L \cdot \frac{L + 3 \cdot L_g}{L + 2 \cdot L_g}$$

$$C_e = 2 \cdot C_p + C_g$$

1.2 Please calculate the equivalent voltage $u_e(t)$ for the simplified case when $L_g \gg L$.

1.3 Assume that the breaker pole of phase R has interrupted at current zero at $t = 0$.

With basis in the single-phase equivalent circuit, derive an expression for the transient recovery voltage $u_b(t)$ over the first phase to clear, i.e. across RR'.

Determine the peak value of the recovery voltage in the first phase to clear when $\omega^2 L_e C_e \ll 1$

In the circuit representation above, the resistances have not been considered. What will the effects on the recovery voltage as a function of time be of including the resistances?

1.4 Assume that the same fault happens in a directly grounded three-phase system (i.e. $L_g = 0$). How large will the peak value of the recovery voltage of the first-phase-to-clear be in this case? Explain briefly.

Problem 2

2.1 Explain briefly how the current injection method can be used to calculate the recovery voltage across a breaker.

2.2 The principle of current injection can also be used in practical measurement for measuring the recovery voltage across a breaker installed in a power system. Describe briefly how this can be achieved in practice. What are the weaknesses and strengths of this method?

2.3 Figure 3.56 shows a circuit breaker that interrupts a large inductive current caused by a terminal fault. Use the principle of current injection

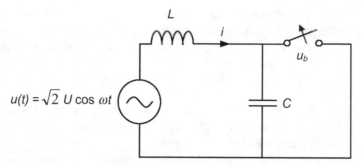

Fig. 3.56 Figure of problem 2

to estimate the recovery voltage across the circuit breaker. The power frequency current flowing through the capacitor C is assumed to be negligible.

2.4 During opening operation, the moving contact of the breaker has a constant speed of 5 m/s. In the first milliseconds after the electric arc is extinguished, the dielectric strength of the contact gap is assumed to be 20 kV/mm. At what time do the contact members have to separate to avoid dielectric re-ignition after the first zero crossing?
Use $L = 30$ mH, $C = 5.3$ µF, $U = 177$ kV and $\omega = 100\pi$.

Problem 3

The circuit diagram in Fig. 3.57 can be used to analyse the interruption of a capacitive current. The capacitive load is given as C_L and is much larger than the network's capacitance to ground C_n. The current interruption takes place at current zero crossing.

3.1 Give some examples of where capacitive current interruption occurs in electric power systems. What are typical current magnitudes?

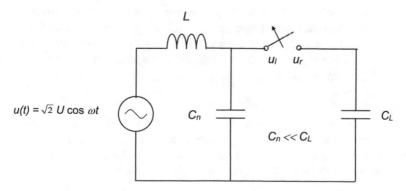

Fig. 3.57 Figure of problem 3

Sketch the waveforms of the current through the circuit breaker and the voltages at the left and right hand sides of the circuit breaker (u_l and u_r respectively) in the time period from a quarter of a power cycle before interruption and until a half cycle after interruption. Assume that the line impedance is negligible compared to the impedance of the capacitive load.

Find an expression for the recovery voltage across the breaker. At what time does the recovery voltage reach its maximum value? How large is this maximum value?

3.2 After the contact members have been separated the movable member has a constant speed of 5 m/s. In the first power cycles after interruption the dielectric strength of the contact gap is assumed to be 15 kV/mm. At what time the contact members have to separate to avoid a re-strike? Assume $U = 300$ kV.

Assume that the breaker re-strikes at the instant when the amplitude of the recovery voltage has its maximum value. Describe how this gradually can lead to over-voltages that are much larger than the source voltage.

3.3 The breaker interrupts at the first current zero crossing. Assume that the impedance due to the inductance L is not negligible compared to the load impedance. (C_n remains much smaller than C_L.)

Establish the network equations for the source side of the system and find an expression for the voltage on the left hand side of the breaker u_l after the interruption. Find also an expression for the voltage at the right hand side of the breaker u_r, and for the recovery voltage across the breaker.

Use $L = 1$ H, $C_L = 2$ µF, $C_n = 0.01$ µF, $U = 300$ kV and sketch the recovery voltage in the first half power cycle after the interruption. Assume that the breaker does not re-strike and that resistive losses result in the transient part being damped out within the first quarter of the power cycle after the interruption.

The dielectric stresses on the breaker are now changed compared to what is the case when the assumptions in 3.1 are valid. What two differences are important with respect to the probability of getting a re-strike? How will the interruption process proceed if the contact separates 1 ms before the current zero crossing and the contact speed and dielectric strength in the gap are as described in 3.2?

Problem 4

What is meant by transient and stationary recovery voltages in the context of current interruption? What normally determines the recovery voltage?

Which two parameters/properties of the recovery voltage are the most decisive ones with regard to how difficult an interruption becomes?

The amplitude of the current that is interrupted also influences on how difficult the interruption becomes. Explain briefly why it is usually more difficult to interrupt a large current than a small one.

Problem 5

5.1 The circuit diagram shown in Fig. 3.58 will be used to examine recovery voltages. Assume that the arc voltage can be neglected and that the power frequency currents flowing through the capacitances C_1 and C_2 are negligible. Interruption takes place at the natural current zero crossing. The voltages at the left and right hand sides of the breaker are u_l and u_r, respectively.
Establish the network equations for the left hand side of the system (source side) and derive an expression for the voltage u_l after the interruption.

5.2 Establish the circuit equations for the right hand side of the breaker (load side) and find an expression for the voltage u_r after the interruption.
Assume that $\omega^2 L_1 C_1 \ll 1$ and that the resonance frequencies at both the source and load side are much higher than the power frequency of the system ω. Show that the recovery voltage just after the interruption can be expressed as:

$$u_b(t) = \sqrt{2}\, U(1 - a_1 \cos \omega_1 t - a_2 \cos \omega_2 t)$$

where a_1 and a_2 are constants determined by the network inductances, and ω_1 and ω_2 are the resonance frequencies of the source and load sides of the network, respectively.

Problem 6

6.1 Figure 3.59 shows a simplified single phase circuit diagram for a 50 Hz grid with an almost entirely inductive load given as $L = 63.7$ mH. The RMS value of the voltage source is $U = 12$ kV. The short circuit inductance $L_{sc} = 1.6$ mH, and the capacitance across the load is $C_l = 4$ nF.

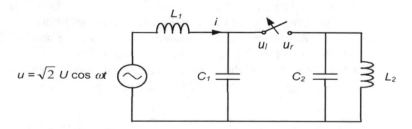

Fig. 3.58 Figure of problem 5

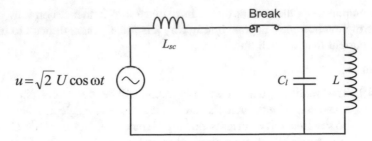

Fig. 3.59 Figure of problem 6

What is the load current, and how large does the stationary short circuit current become in the case of a terminal fault?

6.2 The breaker opens to interrupt a load current. Derive an expression for the recovery voltage when assuming that the current is interrupted at its natural current zero crossing.

Assume here and for the rest of this exercise that arc voltage can be ignored, and that only negligible currents flow through the capacitances before interruption.

Sketch the waveform of the recovery voltage with the given circuit parameters when assuming that the transient part is damped by resistances in the circuit (not shown in the circuit diagram.)

6.3 Assume now that the current is not interrupted at its natural zero crossing, but that the current is chopped at $i_0 = -10$ A just before the zero crossing.

Find an expression of the recovery voltage, and show that it will have an extra term equal to

$$i_0 \sqrt{\frac{L}{C_l}} \sin \frac{t}{\sqrt{LC_l}}$$

Compare to the case without current-chopping in 6.1, when the small delay between the time of chopping and the time of the maximum circuit voltage is neglected.

6.4 As a third case, we will look into the interruption of a terminal fault the circuit above. To get a physically correct circuit diagram a capacitance of $C_n = 4$ μF must be added between the source side and ground.

Derive an expression for the recovery voltage when assuming that interruption occurs at the natural current zero crossing.

Use the given values of the circuit parameters and draw the shape of the recovery voltage when the transients are damped by the resistances in the circuit.

Fig. 3.60 Figure of problem 7

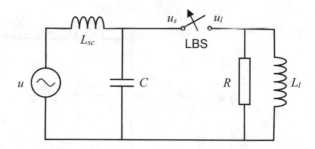

Problem 7

The circuit diagram of Fig. 3.60 is the simplified version of a single phase circuit being used for type testing of medium voltage load break switches.

The supply side consists of a 50 Hz voltage source, a short circuit inductance L_{sc} and a capacitance C. The power frequency current flowing through the capacitance is negligible.

The load consists of a resistance R in parallel with an inductance L_l.

According to the standard, the following conditions should be satisfied:

(i) The supply side should constitute 15% of the total 50 Hz impedance of the circuit

(ii) The load side impedance should have a power factor of $\sqrt{2}/2$

    7.1 A single phase LBS with voltage rating of $U_n = 24$ kV and current rating of $I_n = 630$ A should be tested.

        Determine the values of the components in the test circuit, and show that they become $L_{sc} = 18.2$ mH, $R = 47.9\ \Omega$ and $L_l = 152$ mH.

    7.2 Assume that the arc voltage is negligible compared to the system voltage, and that the current has been interrupted at its natural zero crossing. Derive an expression for the voltage u_l on the load side of the LBS after interruption.

        Explain briefly why the voltage waveform becomes like this; what happens in the load side circuit?

    7.3 Set $C = 0.35\ \mu$F. Determine the frequency of the supply side contribution to the transient recovery voltage.

        If the interruption fails by restrike/re-ignition after around 100–200 μs, which part of the circuit—supply side or load side—is most to blame? Explain briefly (An accurate mathematical analysis is not required; assessments based on semi-quantitative estimates suffice).

Problem 8

The circuit diagram of Fig. 3.61 shows a single-phase power circuit with a resistive load, R. The circuit is supplied by a 50 Hz AC voltage source with the peak voltage U. L is the inductance of the line and C is the capacitance to ground.

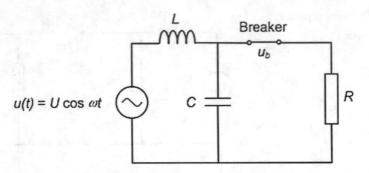

Fig. 3.61 Figure of problem 8

8.1 $L = 6.37$ mH, $R = 3.46$ Ω, $C = 4$ μF and $U = 12$ kV. What is the
stationary load current?
What is the short circuit current in the case of a terminal fault?
What is the ratio between the short circuit current and the load current of
this circuit? How does this compare to a typical power circuit?

8.2 The breaker in the circuit diagram above is requested to open and
interrupt a load current, and by chance the contacts separate exactly as
the source voltage has its maximum. An arc ignites and burns until the
first current zero crossing. Draw the waveforms of the source voltage
and of the current flowing through the circuit in the time interval from
contact separation and till the arc quenches. How long is the arcing time?
Hint: *The phase shift angle ϕ is given by*

$$\tan\phi = \frac{Im\{Z_{load}\}}{Re\{Z_{load}\}} \tag{3.101}$$

*where $Im\{Z_{load}\}$, and $Re\{Z_{load}\}$ are the imaginary and real components
of the load impedance, respectively.*

8.3 Set up the circuit equations and derive an expression for the recovery
voltage, i.e., the voltage that builds up across the breaker contacts after
the arc has extinguished.
Hints: *The capacitance must now be included.*
*Shift the time axis by: $\frac{2}{3}\pi$, so that $t = 0$ corresponds to the time when the
arc is extinguished, that is, let the source voltage be:*

$$u = U \cos\left(\omega t + \frac{2}{3}\pi\right) \tag{3.102}$$

– *The transient solution is assumed to be of the form:*

$$u_t = A \cdot \sin\frac{t}{\sqrt{LC}} + B \cdot \cos\frac{t}{\sqrt{LC}} \qquad (3.103)$$

– *After the current has been interrupted there will be no voltage across the resistive load on the right side of the breaker ($u_r = 0$)*

8.4 (a) What are the frequencies of the transient and stationary parts of the recovery voltage?

(b) What are the maximum amplitudes of the transient and stationary parts of the recovery voltage? Sufficient accuracy is obtained by only including the dominating terms.
Sketch the waveform of the recovery voltage for the first half power cycle assuming that the transient part is damped out within this time interval. Is this a difficult switching duty for the breaker? Explain briefly.

Problem 9

In the circuit shown in Fig. 3.62, a circuit breaker is used for switching of a capacitor bank. If the initial charge of the capacitor bank (C_b) is zero,

9.1 Derive an expression for the inrush current flowing through the capacitor bank, when the circuit breaker is closed at t = 0.

9.2 Discuss the impact of power network frequency (ω) and short circuit current of the network on the amplitude and frequency of the inrush current.

9.3 In the circuit of Fig. 3.62, assume $U = 72$ kV:

- Calculate the short circuit inductance (L_{sc}), if the short circuit current of this network is 40 kA.
- Calculate the capacitance of the capacitor bank (C_b), if the rated current of the capacitor bank is 630 A.
- Consider that a pre-strike occurs by closing operation at a distance of 4 mm in this circuit breaker. If the closing velocity is 1 m/s, calculate

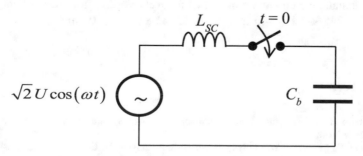

Fig. 3.62 Figure of problem 9

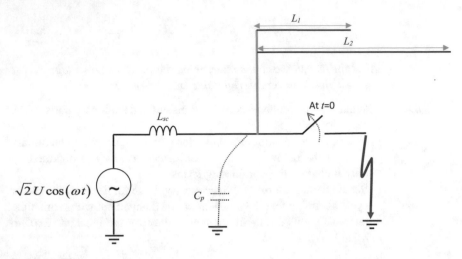

Fig. 3.63 Figure of problem 10

the total energy dissipated in the pre-arc. The arc voltage is constant (300 V).

Problem 10

In Fig. 3.63, a high voltage circuit breaker is used to interrupt a short circuit current caused by a terminal fault. The circuit breaker is connected to a busbar, which is connected to two other transmission lines with wave impedances of R_0 and wave propagation velocity of v. The lengths are L_1 and L_2.

10.1 Please draw a simple equivalent circuit, which can be used for transient recovery voltage calculations for the circuit breaker. Note that, for this purpose, transmission lines are modelled by their wave impedances.

10.2 Derive the governing differential equation for the equivalent circuit with appropriate initial conditions.

10.3 Explain qualitatively how the transient recovery voltage looks like, considering the solution of the simple differential equation and wave propagation along the transmission lines.

10.4 Calculate the initial rate of rise of recovery voltage just after current zero, if $U = 245$ kV, $L_{sc} = 45$ mH, $R_0 = 400~\Omega$ and $C_p = 100$ nF.

References

1. Blackburn JL (1993) Symmetrical components for power system engineering. Marcel Dekker Inc., New York
2. Grainger JJ, Stevenson WD (1994) Power system analysis. McGraw-Hill Inc., New York

3. IEC 62217-100 (2008) High voltage switchgear and controlgear, part 100: alternative current circuit breakers
4. IEC 62271-103 (2011) High voltage switchgear and controlgear, part 103: switches for rated voltages above 1 kV up to and including 52 kV
5. Jonsson E (2014) Load current interruption in air for medium voltage ratings. Doctoral thesis 2014:83, Norwegian University of Science and Technology (NTNU)
6. IEEE standard C37.083 (1999) IEEE guide for synthetic capacitive current switching tests
7. CIGRÉ WG 13.02 (1980) Interruption of small inductive currents. Electra 72:73–103

Chapter 4
Current Interruption Technologies

In the preceding chapters, the basics of current interruption in power switching devices with mechanically opening contacts have been reviewed. Furthermore, different scenarios of application of switching devices in power networks were thoroughly discussed. In this chapter, we aim to present the technologies currently used to realize the switching components. As explained in Chap. 1, a switchgear has different parts such as interruption chamber and drive mechanism, and it may be used for different switching operations. In the first part of the present chapter, different types of switching devices are introduced based on their functionalities and applications.

In power networks, switching components are usually installed in power substations. Hence, it is important to review the basics of the high voltage substations to become familiar with different types of components used to realize the switching operations.

These include disconnector switches, load break switches, earthing switches, fuses and circuit breakers. The latter sections of this chapter cover the technologies used to realize those functionalities in power switching components.

4.1 High Voltage Substations

4.1.1 Introduction

Switching devices are quite often located together with power transformers. The transformers may be used to step down the transmission system voltage to distribution level voltages for electricity supply in the region, or to step up the generator voltage from a nearby power generating station to transmission system levels. Substations for transmission system voltage levels (145–800 kV) typically link 3–10 power lines/transformers. In addition, the substation may include reactors and

© Springer International Publishing AG 2017
K. Niayesh and M. Runde, *Power Switching Components*, Power Systems,
DOI 10.1007/978-3-319-51460-4_4

capacitor banks for reactive power compensation, and also other components. Each link is normally referred to as a *branch*.

The number of substations is of course much higher at lower voltage levels and in industrial networks. The voltage levels used in distribution systems are typically a few tens of kilovolts; in particular is equipment rated for 24 kV widely used. Distribution level substations are also quite often located together with transformers. The size, rating and complexity of distribution substations vary greatly, from a very few branches in sparsely populated areas to installation of several tens of branches in the larger cities or in large industrial plants.

4.1.2   Configurations

A substation consists of a number of high voltage components installed close to each other and connected with copper or aluminium conductors. Several types of current interrupting devices (circuit breakers, fuses, load break switches, disconnector switches) are used to change the grid configuration and/or to disconnect faulty parts. Earthing switches are used to ground components when maintenance or repair work is being carried out. Current and voltage transformers are used for metering and protection purposes; and surge arrestors for protecting against overvoltages.

There exists a large number of ways of connecting the different components into a complete substation. The chosen solution is typically a result of a trade-off between redundancy requirements (need for back up if some component fails and must be taken out of service) and total costs.

A common configuration or "architecture", at least at the highest voltage levels, is to use what is usually referred to as double busbar/two-breaker system. Such a substation layout in a case with three branches, two transmission lines and one transformer branch, is shown in Fig. 4.1.

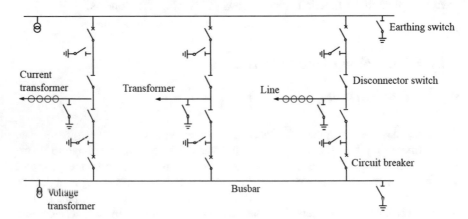

Fig. 4.1 A double busbar/two-breaker substation with three branches (*single line* diagram). All breakers and earthing switches are drawn in open position

A two-breaker system means that there are two circuit breakers per branch. The double busbar/two-breaker system has a high degree of redundancy. If a fault occurs or maintenance is to be carried out, many of the substation components can be taken out of service, but all branches of the substation can still be operative. For example, by opening the disconnector switches between the circuit breakers and the branches on one busbar, the busbar and the three circuit breakers can be taken out of service and grounded while all branches are operative by using the other half of the substation. The only components where there is no redundancy are the first earthing switch and the current transformer closest to the incoming line.

The double busbar/two-breaker arrangement is a simple and easy-to-follow configuration. The great disadvantage is of course the large cost. The many circuit breakers contribute significantly to the total cost.

An alternative arrangement for the same three-branch substation is shown in Fig. 4.2. Here, there is only one busbar and one circuit breaker per branch. The number of circuit breakers is halved from six to three; the number of disconnector switches is reduced from 17 to 13. Consequently, such a single busbar/one-breaker configuration is a far less expensive solution. The disadvantages are just as obvious; the substation is very vulnerable in case of component failures, and maintenance becomes more complicated. Nearly any fault or maintenance operation renders at least one branch out of service.

In addition to the configurations shown in Figs. 4.1 and 4.2, several other ways exist to arrange the components into a complete substation [1]. There are substations with three busbars, with by-pass circuit breakers, with busbars that can be split, with many disconnector switches and relatively few circuit breakers, etc. The cost and redundancy for these solutions are somewhere between the two architectures or configurations just described [2].

For transmission level substations, which are of vital importance to the overall system reliability, the double busbars/two-breaker arrangement is quite common.

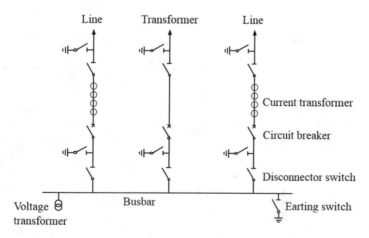

Fig. 4.2 Single busbar/one-breaker substation (*single line* diagram)

For less important sub-transmission level substations, and distribution and industry substations, solutions somewhat more redundant than the arrangement of Fig. 4.2, but less extensive and expensive than that of Fig. 4.1 are typically used. High voltage components have in general become increasingly reliable and economic criteria appear to have become more important than earlier. The tendency is thus to choose substation configurations with less circuit breakers redundancy.

As seen in Figs. 4.1 and 4.2, the components used for switching operations at the highest voltage levels (transmission or sub-transmission substations) are circuit breakers, disconnector switches and earthing switches.

In distribution networks, in addition to single or double busbar configurations as in transmission level substations, it is possible to have the so-called ring configuration as shown in Fig. 4.3, where many simple substations, referred to as *ring main units* (*RMU*), are connected to the adjacent switching substations from both sides [3]. In such configurations, the functions of load switching and short circuit interruption are performed by two different components. Load break switches are used to energize and de-energize a load or part of a network, and short circuit currents are interrupted by either a circuit breaker or a combination of load break switch and fuse, placed on the outgoing feeder.

The ring circuit is normally split in two parts (as shown in Fig. 4.3). So, some distribution substations are energized through one feeder and some other through another feeder. In case of a fault, operation of the corresponding circuit breaker of

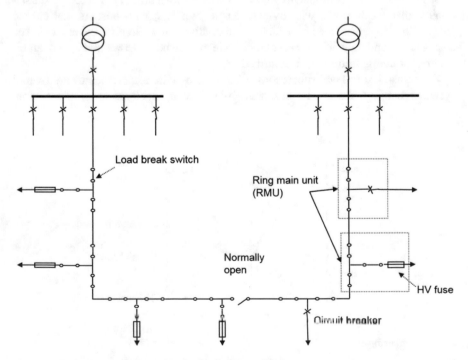

Fig. 4.3 Typical ring configuration in distribution networks

the upstream substation and re-arrangement of the load break switches isolate the faulty part of the ring and re-establishe the energy supply to the remaining healthy part of the ring.

Two different technologies for providing electric insulation to substations are being applied. In *air insulated substations* (AIS) atmospheric air provides the insulation between phases and to ground (in addition to solid insulation like porcelain). In *gas insulated substations* (GIS) on the other hand, all components are placed inside gas tight enclosures (filled with SF_6 or other pressurized gases), so here the pressurized gas is the main insulation (in addition to solid insulation like epoxy). In recent years, equipment combining elements from both AIS and GIS has become available. This is referred to as *mixed technology substations*. It thus seems likely that the very clear difference that has existed between the AIS and GIS technologies is becoming less pronounced. The architecture or configuration of a substation is largely independent of the used technology.

4.1.3   Air Insulated Substations (AIS)

In air insulated substations, the high voltage components are erected individually and connected with bare metal conductors. Normal atmospheric air is thus used as electric insulation in addition to the insulation (porcelain, epoxy, oil-impregnated paper, SF_6 etc.) of the individual components. The large distances associated with air insulation mean that substations for the highest voltages require much space. Hence, they are normally located outdoor. Figure 4.4 shows a typical outdoor 145 kV substation.

The individual components in AIS are connected with normal aluminium overhead line conductors, while the busbars often are made of thick-walled aluminium tubes. At distribution level voltages, the insulating distances are shorter and the substations are usually located indoor. Distribution and industry grid substations are often prefabricated units containing several components. Each unit is mounted and then connected to the busbar.

The individual components are easily accessible in air insulated substations, see the example in Fig. 4.4. It is relatively simple to locate and access a failed component and to perform necessary repairs and replacements. The substation is also easy to inspect visually. For example, whether a disconnector switch or earthing switch is closed or open can be observed directly.

An important and obvious drawback with AIS is the large insulating distances, meaning air insulated substations for the highest voltage levels demand a lot of space. A 420 kV substation of the configuration shown in Fig. 4.1 covers several thousand square meters per branch. Moreover, outdoor substations are also much more exposed to the environmental stresses than indoor substations. As the lifetime can be as long as 40 and 50 years, the components have to be carefully designed with regard to environmental factors such as rain, snow, fog, cold climate, pollution and ultra violet radiation.

Fig. 4.4 Part of a 145 kV air insulated substation. The row of components to the *left* are SF$_6$ circuit breakers, the row to the *right* are disconnector switches with air as the interruption medium. Two busbars are visible in the rear to the *left*

4.1.4 Gas Insulated Substations (GIS)

Gas insulated substations came into use around 1970 and became a competing technology to the air insulated substations, which until then were alone on the market.

GIS technology is available for system voltages of 72.5 or 145 kV and above. All substation components are mounted inside earthed steel or aluminium enclosures or ducts. These are filled with SF$_6$ at 5–8 bars pressure. SF$_6$ is thus used as both interruption medium in switchgear and as insulating medium in the entire substation. In addition, epoxy is used as solid insulation. One of the world largest gas insulated substations installed at Three Gorges Dam in China is shown in Fig. 4.5.

The high voltage lines are fed through the walls of the building by means of large SF$_6$ filled porcelain bushings (not in the photo). Each phase has its own duct.

The ducts from the branches are then led down and split in one conductor to the right and one to the left. This substation has a two-breaker/double busbar configuration (like the arrangement shown in Fig. 4.1) and 73 branches. The circuit breakers with two interrupting chambers are mounted horizontally inside the large cylindrical SF$_6$ filled compartments seen in the lower left half of the photo, whereas

Fig. 4.5 A 550 kV SF$_6$ gas insulated switchgear installation, at Three Gorges Dam in China with 73 bays (Courtesy of ABB)

a few of the circuit breakers on the other busbar are barely visible at the back. The circuit breakers are in the same design as those in air insulated SF$_6$ switchgear. The circuit breakers have hydro-mechanical spring operating mechanisms, which are located outside the enclosure.

This GIS has single-phase encapsulation, i.e. each phase conductor and all components on each phase have separate ducts. There are also gas-insulated substations with three-phase encapsulation, but this is more common at lower voltages, typically 145 kV. Figure 4.6 shows schematically the conductor configuration in a single-phase and three-phase enclosure, respectively.

Usually, thick-walled aluminium cylinders or tubes are used as high voltage conductors. Complete gas insulated substations are normally erected by assembling smaller units or "building blocks". Hence, there are many contacts/joints along the high voltage conductor and many flanges with o-ring seals in the enclosure.

Epoxy insulators inside the encapsulation are used to keep the high voltage conductors in place. Most often, these are bell-shaped cones or "cups" as such a shape somewhat reduces the dielectric stresses they are subjected to. An epoxy insulator from a single-phase enclosure is shown in Fig. 4.7.

Figure 4.8 shows schematically how the high voltage conductor is held up and supported by such insulators in a single-phase enclosed section.

A GIS is divided into several separate gas compartments. This is in order to limit the SF$_6$ emissions if there is a leakage, or if a powerful arcing failure burns through

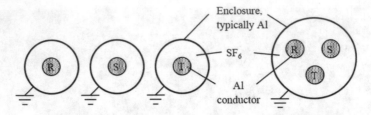

Fig. 4.6 Conductor configuration in single-phase (*left*) and three-phase (*right*) GIS encapsulations

Fig. 4.7 Epoxy insulator with a metal conductor and ring contact in the *centre*

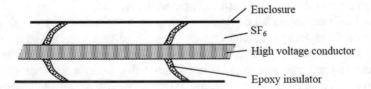

Fig. 4.8 Cross section of a GIS conductor section of a single-phase enclosure

the encapsulation. The epoxy insulators can be made gas tight and have a function in separating the different gas compartments.

The good insulating properties of the SF_6 make it possible to design very compact gas insulating substations. This is the most important advantage of this technology. While insulation distances of several meters are required in transmission level substations when relying on atmospheric air for insulation, only 10–20 cm is sufficient if the gap is filled with SF_6 at 5–6 bars. A GIS typically needs only about 10% of the space (volume) required by a corresponding AIS.

The encapsulation contributes significantly to the cost and makes GIS more expensive than AIS, at least when only considering the costs of high voltage components. When also taking cost of land, buildings, maintenance etc. into account, the picture becomes more complex, and a general cost comparison is impossible.

In some countries, GISs are always located indoor, while others have large populations of GIS installed outdoor. GIS is mainly used where high property costs (in cities) or lack of space (such as in hydroelectric power plants in man-made caverns) justifies the higher component costs. Also at places where harsh environmental impacts make outdoor air insulated substations unsuitable, e.g. in the high mountains where bad weather and a lots of snow make the access to the substation difficult during the winter.

The encapsulation is always grounded. Hence, with regard to personnel safety GIS are considered better than AIS.

The failure rates associated with GIS have in general been found to be somewhat lower than for AIS. However, experience has shown that for severe failures, typically those involving internal arcs, repairs are complicated, expensive and time consuming. The SF_6 gas must be evacuated and the enclosure must be opened before the component can be accessed. Consequently, the duration at which a gas insulated substation is out of service after a failure is often considerably longer than for a similar type of failure in an air insulated substation. The costs of repair are also in average considerably higher. Repair times of several weeks and even months after severe failures are rather common. This is an obvious disadvantage with gas insulated substations.

4.1.5 Distribution Level Substations

Technologies with elements from both AIS and GIS are also used at medium voltage/distribution level substations, both in so-called primary and in secondary distribution substations. Secondary distribution substations are simpler and often without circuit breakers; a load break switch in series with a fuse is used instead. The substation design and arrangements also show greater variety. The required insulation distances are rather small compared to high voltage equipment; this makes it possible to fit different apparatus of a distribution switchgear into a cabinet/cubical.

Such switchgear cubicals are often divided in a number of different compartments, namely circuit breaker compartment, busbar compartment, (cable) connection compartment and low voltage compartment. These compartments may be completely isolated from each other with solid metallic walls (metal clad). Another alternative is to have some parts together in one place (metal enclosed). It is quite common that some parts of a substation, e.g. circuit breakers and busbars, are in a gas tight enclosure. Both metal clad and metal enclosed designs may be air or gas insulated. An exemplary metal-clad air insulated medium voltage switchgear is shown in Fig. 4.9.

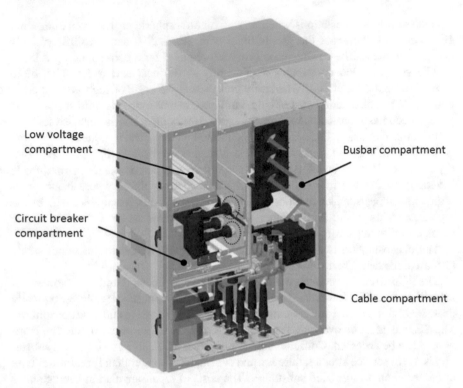

Fig. 4.9 Exemplary air-insulated metal clad medium voltage switchgear (Courtesy of ABB)

In many such switchgears, the circuit breaker is withdrawable. The connecting points of the circuit breaker to the other compartments (dotted circles in Fig. 4.9) fulfil the function of disconnecting switches, which isolate the circuit breaker when a repair/maintenance operation is performed on the circuit breaker.

4.2   Interrupting Media

The term interrupting medium refers to the medium, usually a gas, in which the arc burns. As an introduction to the different interrupting media, it may be interesting to list the properties of a "perfect" or "ideal" interrupting medium.

4.2.1   Desired Properties

The tasks of the arc vary substantially over the different phases of the arcing period during a current interruption process. Hence, the required properties of the medium vary accordingly.

When *high current* flows through the arc, it is important to limit the energy dissipated in the switchgear. Since the dissipated energy is proportional to the arc voltage, the arc voltage should be small when the current is large. The arc voltage is directly related to the electric conductivity of the arc. A good electric conductivity is obtained when the plasma contains many and fast charge carriers, preferably electrons. A large charge carrier density is achieved when the ionisation energy is low and/or the arc temperature is high, see Fig. 2.11. Moreover, maintaining a high arc temperature is easier with a low thermal conductivity. Hence, at high currents, *low ionisation energy* and *low thermal conductivity* are desirable.

At *current zero crossing* a fast transition from conducting to insulating state is required, meaning a small time constant for this process. This implies a rapid cooling of the arc, which is easier to achieve if the thermal conductivity is high at relatively low temperatures. Large ionisation energy, resulting in a low density of free electrons, is also beneficial at these temperatures. Hence, at low currents close to current zero crossing a *high ionisation energy* and a *high thermal conductivity* are desired.

After *current zero*, it is important to quickly establish a high dielectric strength in the gap. This is achieved either by reducing the charge carrier density through a fast recombination to neutral particles, or by attaching free electrons to neutral particles and thereby producing heavy and low mobility negative ions. The latter process will take great advantage of a medium with *good electronegative properties*.

In the special case of current limiting switchgears and for DC breakers, completely different properties are desired. A medium that has poor electric conductivity also at large currents is here preferred. This is achieved with a high ionisation energy and a high thermal conductivity.

In addition to the desired arcing properties, many other aspects must be taken into consideration when assessing the suitability of an interruption medium. These include corrosivity, toxicity and other environmental aspects. The substance must be chemically stable for several decades, and the price should be low. When using a molecular gas such as SF_6, it is also essential that it, after decomposing in the hot arc, recombines when cooled down.

4.2.2 Sulphur Hexafluoride (SF_6)

SF_6 is an electronegative gas with good dielectric properties and extremely good arc quenching properties. The physical explanations for its good interruption properties are outlined in the following.

SF_6 dissociates (decomposes) at relatively low temperatures, about 1000–4000 K. Consequently, the properties at large currents and high temperatures are essentially determined by the properties of the dissociation products, i.e. sulphur and fluorine. The ionisation energies of S and F are 10.4 and 17.4 eV, respectively, as shown in Table 2.1. The value for sulphur is relatively low, leading to good electric conductivity at high currents. Thus, the arc voltage and energy dissipation are low.

At low currents (close to current zero crossing) the input power reduces, causing the arc temperature to decrease. The dissociation products then recombine and form SF_6 again. This leads to a decreasing electric conductivity, as the ionisation energy of SF_6 is 19.3 eV, which is considerably higher than for sulphur.

The thermal conductivity is relatively good in the 1000–3000 K range. Due to these two properties (high ionisation energy and good thermal conductivity), the arc time constant close to current zero crossing becomes extremely low; in the microsecond range (see Table 2.2).

After the arc is extinguished at current zero crossing, the electronegative properties of the SF_6 gas cause any free electrons in the gap to combine with neutral particles and form slow negative ions. Consequently, a high dielectric strength quickly builds up between the contacts.

The significant change in ionisation energy at current zero crossing, and the very good thermal conductivity at temperatures *below* the ionisation temperatures of the molecular gas are thus the most important reasons for the excellent interruption properties of SF_6.

SF_6 is almost an inert gas. It is chemically very stable, non-corrosive and non-toxic. However, toxic and corrosive dissociation products form when an arc is burning in SF_6. When the arc is extinguished, practically all the gas recombines, but this small fraction may create very reactive compounds (e.g. hydrofluoric acid) if moisture is present. Consequently, SF_6 switchgears are typically equipped with a filter with a substance that attracts both moisture and the SF_6 dissociation products. A frequently used filter material is activated aluminium oxide.

Interaction between the arc and metal parts, e.g. the contacts, can lead to formation of metal fluorides, which appear as a fine, light grey powder inside the interrupting chamber [4]. These metal fluorides are electric insulators and do not cause dielectric problems. In gases containing carbon, similar phenomena may impair the insulating properties due to the formation of electrically conducting carbon compounds.

The very good properties of SF_6 around current zero crossing have made it the best interruption medium found. SF_6 at 4–8 bar is the totally dominating interruption medium in modern circuit breakers for voltages exceeding 100 kV. SF_6 is also widely applied in switchgears for lower ratings, for example in distribution and industry systems, then at a pressure of 1–1.3 bar.

SF_6 efficiently absorbs infrared radiation, and since it decomposes very slowly when released to the atmosphere, it is a very potent "greenhouse gas". These unfortunate environmental aspects are the greatest drawback associated with using SF_6 as interruption and insulation medium.

SF_6 have had a rather modest price. However, lately there has been a considerable cost increase. It is also reasonable to assume that the environmental aspects will cause governments and other authorities to seek to reduce the use and emissions of SF_6 through restrictions, taxes and other impositions. These may be rather dramatic and over a period of some years, they may budge the very dominating role of SF_6 as an interruption medium.

4.2.3   Air/Nitrogen

Air contains nearly 80% nitrogen. Hence, when considering air as an interruption medium, the characteristic data for N_2 are the most interesting.

The dissociation energy for N_2 is somewhat higher than for SF_6, so nitrogen decomposes in the temperature range of 4000–10,000 K, see Fig. 2.10. The ionisation starts at these temperatures and the gas changes from being insulating to conducting over approximately the same temperature range. Thus, at high currents, the electric conductivity is good, and consequently, the arc voltage is low, but not as low as for SF_6.

At lower currents (close to current zero crossing) the ionisation energy for the nitrogen molecule is larger than for the free atoms. The difference is not as pronounced as between SF_6 and S, but the transition from conducting to insulating state also for nitrogen becomes sharper since the ionisation energy of the molecule is greater than that of the atoms.

The maximum thermal conductivity for nitrogen occurs at a relatively high temperature compared to SF_6. It coincides with the temperature range in which the transition from insulating to electric conducting state occurs (see Figs. 2.12 and 4.10).

Its thermal conductivity is rather poor for temperatures below 3000 K, and the dielectric strength is not fully recovered at these temperatures. For this, but also other reasons, the time constant of an arc burning in air becomes considerably larger than in SF_6.

Nevertheless, the physical properties that are important in the present context are mainly good, and both nitrogen and air are very well suited interruption media compared to most other gases. Nitrogen and air have several other important advantages; they are non-toxic, incombustible, non-corrosive and inexpensive, and have definitely no negative environmental properties.

In many switchgears for lower ratings, the arc chamber is not enclosed so the arc is burning in regular atmospheric air. Air is also used in circuit breakers of higher rating, then normally at higher pressures.

4.2.4   Oil

An arc burning in oil is in reality burning in a gas mixture formed by thermal decomposing of the hydrocarbons of the oil. The gas mixture consists mainly of

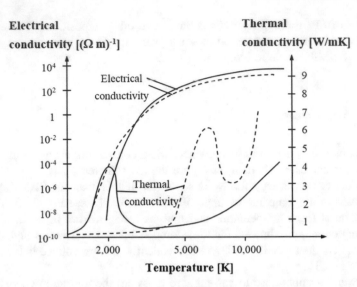

Fig. 4.10 Electrical and thermal conductivity for SF$_6$ (*solid*) and nitrogen (*dashed*)

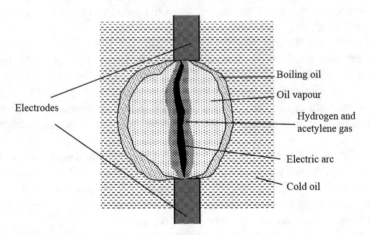

Fig. 4.11 An electric arc burning inside a "gas bubble" between the electrodes in an oil filled switchgear

hydrogen (70–80%) and acetylene (15–20%), but does also contain smaller fractions of methane and other gases. A schematic drawing of an arc burning in oil is shown in Fig. 4.11.

The arc temperature is 5000–15,000 K, and it is completely surrounded by dissociated gas at nearly the same temperature. Outside this, there is a volume of vaporised oil at considerably lower temperature. Then there is an interface of boiling oil towards the bulk oil volume.

Since the arc in oil-filled breakers is also burning in a gas, the descriptions and explanations given above for gas-filled switchgear are largely valid for oil-filled breakers.

The very high thermal conductivity over a large temperature range for hydrogen (see Fig. 2.12) is due to its low atomic weight. At relatively large currents, this is somewhat disadvantageous as the arc temperature becomes lower than when SF_6 or N_2 is used, and the arc voltage is correspondingly higher. Close to the current zero crossing, the high thermal conductivity leads to a rapid transition from conducting to insulating state. Hence, the arc time constant for an arc burning in hydrogen is very small (see Table 2.2). This is the single most important reason for hydrogen being such a good interrupting medium.

Small quantities of carbon particles are formed when the oil is decomposed by the arc. After a number of switching operations, the oil can be so contaminated by the particles that the dielectric strength is reduced. The oil must then be cleaned or replaced. Compared to switchgears with SF_6 or nitrogen as the interruption medium, oil filled breakers need considerably more maintenance.

At first sight, it may look rather daring to use a liquid as flammable as oil in a switchgear. However, it works out fine as long as the generated gases do not get into contact with air while the arc is burning. If the interruption chamber is ruptured during a switching operation, or if the gases formed by the arc in some other way come into contact with air and an ignition source, the danger of having an explosion is imminent. Quite a few such accidents with oil-filled breakers have occurred, and the fire hazard is definitely a disadvantage associated with this arcing medium.

4.2.5 Vacuum

Besides high pressure switching arc, which is used in oil and gas switchgear, it is also possible to have very low-pressure switching devices. These are referred to as vacuum switchgear.

Very good insulation property in vacuum due to lack of ionisable gas atoms/molecules was the first motivation to use very low-pressure environment to realise high voltage switching devices. Later, it was understood that in very low-pressure environments breakdown initiates from the electrodes and the ionisation processes in the vacuum volume play a subordinate role [5]. Hence, the breakdown voltage is not linearly dependent on the gap length and shows a non-linear, saturating behaviour as shown schematically in Fig. 4.12. Consequently, the application of vacuum switching devices to the medium voltage range, with maximum impulse withstand voltage levels of about 200 kV, results in very compact and cost-competitive solutions.

As explained in Chap. 2, the nature of the switching arc in vacuum switchgear is very different compared to the other types, as there is almost no background gas, which can be ionized. The necessary charge carriers are mainly provided from the contact surfaces by different emission processes. Near current zero crossing, if no

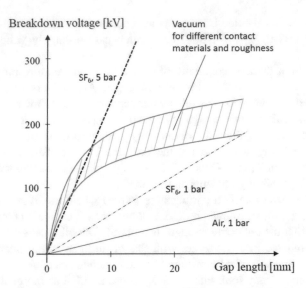

Fig. 4.12 Voltage withstand of a vacuum gap for different gap lengths compared to other insulation materials

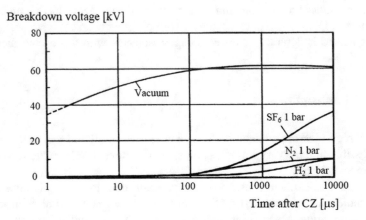

Fig. 4.13 Dielectric recovery of a vacuum switching gap compared to other interrupting media

new charge carriers are released to the switching gap, the available charge carriers diffuse out of the gap, which results in a very fast dielectric recovery as shown in Fig. 4.13. This is the case when there are no large and deep molten regions formed on the contact surfaces at the time of current zero. In Fig. 4.13, the breakdown voltage of a 6 mm gap after interruption of a 1600 A current is shown for different interrupting media [6].

Under these circumstances, after the arc is extinguished, the dielectric strength in the gap increases rapidly due to a swift disappearance of the ions and metal vapour from the gap. The ion drift velocity is about 10^4 m/s, and the gap between the contacts is virtually without charge carriers within a couple of microseconds.

To avoid large molten regions on the contacts, it is important to make the energy flux to the electrodes as homogeneous as possible. In conventional vacuum switching devices, this is achieved by applying either radial or axial magnetic fields to the vacuum arc. The required magnetic fields are generated either by current flow through the appropriately shaped electrodes or using external magnetic coils (see Sect. 4.4.5).

4.3 Transmission Level Circuit Breakers

4.3.1 Technologies

As mentioned in Chap. 1, a circuit breaker is supposed to interrupt all currents and voltages that can occur in the position in the grid where the breaker is installed. This makes the circuit breakers the most important and expensive switchgear type.

Presently, there are four circuit breaker technologies in use at transmission level voltages: SF_6, minimum oil, bulk oil and air blast circuit breakers. The distribution between the different technologies is illustrated in Fig. 4.14.

What also can be seen in this figure is the distinct development with regard to the preferred circuit breaker technology. The air blast circuit breaker part around 1960 was more than 50%, but decreased as minimum oil breakers took over the entire market for new installations in the 60-ies and 70-ies. Around 1970, the first SF_6 circuit breakers were put into service. From then on, this technology totally dominated the market. There were practically no more minimum oil breakers in new installations. Single pressure SF_6 technology has been completely dominating the high voltage switchgear since mid 80-ies in most markets.

Most of the circuit breakers for theses voltage levels only operate around 10–20 times a year, and the wear is therefore insignificant so they last for decades.

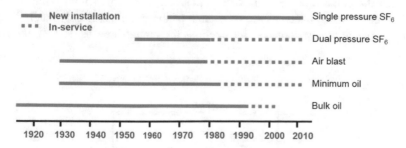

Fig. 4.14 Transmission level circuit breaker technologies for in-service and new installations

Consequently, many of the circuit breakers installed in the early 60-ies are still in service.

The most frequent reason for replacing a circuit breaker at these voltage levels is when its short-circuit current rating (the maximum current it can interrupt) no longer is sufficient, not because the switchgear is worn-out. Extensive developments of the grid necessarily lead to greater short-circuit currents, and this causes some breakers in otherwise good condition to be replaced. Nevertheless, a service life of 40 years is not uncommon.

Based on application of different interrupting media, different types of high voltage circuit breakers have been developed. The switching arc in most of the media of Fig. 4.14 is of the high pressure type. As described in Chap. 2, a successful current interruption in high pressure switching arcs can only occur if the arc is efficiently cooled. This implies that the arc in the circuit breaker is almost always in a fixed position, where the gas flow blowing on the arc is optimized. In many circuit breakers, this is realized by using a pair of contacts, which are enclosed with an axisymmetric dielectric body, which is referred to as the *nozzle*. The diameter of the contacts is as small as possible to maximize the gas velocity, but they have to be able to carry the short circuit current during the current interruption without any permanent damage. Such small diameter contacts, which are exposed to the switching arc and made of materials with rather high resistivity, cannot provide sufficiently low contact resistances in the closed position.

Therefore, the tasks of current carrying and current interruption are normally separated by using at least two pairs of contacts, namely main contacts (responsible for current carrying in the closed position) and arcing contacts (between them the switching arc burns and current interruption takes place). By opening operation of a circuit breaker, first the main contacts separate; the current is then commutated to the arcing contacts and afterwards by separation of arcing contacts, the switching arc burns between arcing contacts.

The process of current commutation can be understood considering a simple equivalent circuit of the contact systems of a circuit breaker, see Fig. 4.15, where S_1 represents the main contacts and S_2 the arcing contacts. R and L represent the resistance of the arcing contacts and the total inductance of the loop between main and arcing contacts, respectively.

When the circuit breaker is in closed position, the current flows almost solely through the main contacts, as the resistance of the main contacts is much smaller than that of the arcing contacts. By opening the circuit breaker, at first, the main

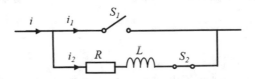

Fig. 4.15 Equivalent of contact systems of a circuit breaker during the current commutation phase

contacts are separated and an arc ignites between them. This arc has to be extinguished very quickly by commutation of current to the parallel RL path.

The governing equations for the current commutation phase may be expressed as follows:

$$\begin{cases} i_1 + i_2 = I_m \cos(\omega t + \varphi) \\ Ri_2 + L\frac{di_2}{dt} = u_{arc}(i_1) \end{cases} \quad t > 0$$
$$i_1(0^-) = I_m \cos \varphi, \quad i_2(0^-) = 0 \tag{4.1}$$

Here u_{arc} is the voltage drop of the arc burning between the main contacts and the total current i is assumed to be sinusoidal with the frequency of ω and the amplitude of I_m. The separation of the main contacts happens at the phase angel φ.

In a simplified case, when the arc voltage is constant and equal to U_0, the current flowing through the arcing contacts can be expressed as:

$$i_2(t) = \frac{U_0}{R}\left(1 - e^{-\frac{R}{L}t}\right) \tag{4.2}$$

Current commutation is completed when the current of the arcing contacts becomes equal to the total current flowing through the circuit breaker. The maximum current commutation time corresponds to the maximum simultaneous current (i.e. I_m) and can be calculated as:

$$t_{commutation} = -\frac{L}{R}\ln\left(1 - \frac{R \cdot I_m}{U_0}\right) \tag{4.3}$$

To ensure a successful current commutation to the arcing contacts, the arcing contacts have to be separated with a delay in respect to the main contacts. This delay must be larger than the maximum current commutation time, given by (4.3).

In the following sections, it has been assumed that a successful current commutation to the arcing contacts occurred and therefore the focus is on the processes related to the current interruption between the arcing contacts.

4.3.2   Air Blast Circuit Breakers

Air blast circuit breakers utilize air at high pressure, typically 20–40 bar, as insulating medium and interrupting medium as well as to power a pneumatic operating mechanism [7]. During current interruption, the contacts are quickly opened and separated, and at the same time a valve opens and a high-pressure air blast cools the arc and prevents it from re-igniting after the current zero crossing.

In order to achieve the correct and efficient air flow on the burning arc, contact and nozzle design are very important. Some of the most popular solutions are shown in Fig. 4.16. The contacts and nozzles are typically concentric and normally encapsulated in a hollow, cylindrical shaped porcelain insulator (not included in the drawing) with metal terminations at both ends.

The nozzles are often but not always, made of insulating material and their task is to make sure that the air flow blows efficiently on the arc while the contacts separate. The nozzles can be fixed, or the switchgear can be designed in such a way that they completely or partly follow the axial movement of the contacts during the switching operations.

Axial or radial flow is most common for circuit breakers with very high ratings. The *double flow nozzle* type in the middle of Fig. 4.16 is a particularly successful design as the air flow is split and blows in both directions.

Air blast circuit breakers for the highest voltage ratings are designed using a series combination of several interrupter units or extinguishing chambers. Both six and eight in series, installed in separate porcelain insulators are common for the highest ratings. The stresses due to the recovery voltage are thus, at least in theory, divided equally between the different chambers. Voltage grading capacitors in parallel with each interrupter is used in order to achieve a more uniform division of the voltage stresses during interruption, see Fig. 4.17. The grading capacitors are generally mounted in separate housings.

The purpose of the voltage grading capacitors is to reduce the impact of stray capacitances to earth. If the grading capacitors are much larger than the stray capacitances, the impact of the stray capacitances can be neglected. In this case, the grading capacitances act as a capacitive voltage divider during switching transients and when the contacts are in open position. If the capacitors have the same capacitance, the dielectric stresses are split equally between all the interrupters.

Compressed air is a good electric insulator; at a pressure of 20 bar, the dielectric strength is about the same as for oil. Consequently, air blast circuit breakers were often designed to maintain high pressure in the interruption chamber also after

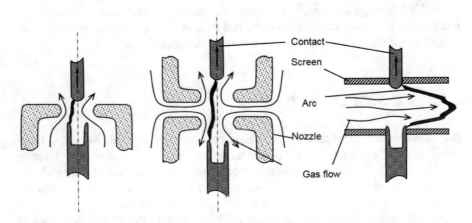

Fig. 4.16 Different nozzle and contact configurations during interruption in air blast circuit breakers; axial blast (*left*), radial blast (*middle*), and lateral blast (*right*)

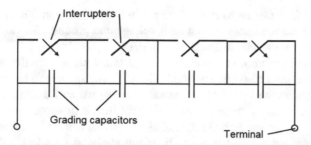

Fig. 4.17 Interrupter units and grading capacitors in a circuit breaker with four arcing chambers in series (*schematic*)

interruptions, i.e. when the circuit breaker is in open position. The obvious disadvantage is that an air leakage can considerably reduce the dielectric strength across an open breaker.

An alternative solution to permanently maintain a high pressure in the interruption chamber is to use a disconnector switch in series with the interruption chamber. The disconnector opens some tens of milliseconds after the circuit breaker contacts have opened, increasing the dielectric strength of the breaker.

During opening operations, air blast circuit breakers vent the compressed air that has cooled the arc. This causes intense noise and is a clear disadvantage of this circuit breaker technology. Noise problems have in many cases been an important reason for replacing air blast circuit breakers.

The first air blast circuit breakers were developed in the 40-ies and this technology was dominating in the 50-ies. For a while, the air blast circuit breaker was the only commercially available option for voltages above 300 kV. Several interruption chambers in series, high pressure and large mechanical stresses in the interruption chamber, the need for disconnector switches, and large compressors make air blast circuit breakers complicated and expensive. The maintenance is also rather extensive. In particular, it has turned out to be difficult to sustain a sufficiently high reliability of compressors and other parts of the compressed air system when dealing with pressures up to 150 bar, especially when much of the equipment are installed outdoor.

As previously mentioned, air blast circuit breakers lost market share to minimum oil breakers in the 1960-ies. Presently, only few remain in service.

As generator circuit breakers, where the voltages are lower and the currents are higher than in the transmission grid (typically $U_N = 10$–20 kV/$I_N = 5000$–$10,000$ A), air blast circuit breakers are still quite common.

4.3.3   Minimum Oil Circuit Breakers

Two technologies rely on oil as the interruption medium: *Bulk oil* (*oil tank*) and *minimum oil* breakers. The most important difference is that minimum oil breakers use oil only as interruption medium, whereas in bulk oil (oil tank) breakers, it is also used as insulation medium between high voltage and ground. The designs of these two types are shown schematically in Fig. 4.18.

The oil tank breaker looks very much like a transformer from the outside. It is a large, grounded steel case filled with oil. The high voltage line to be switched is fed in and out of the case by porcelain bushing insulators, and the interrupter is submerged in oil. Some of these oil tanks are still in use in some countries, demonstrating that they are sturdy and reliable, although this is a totally out-dated design. As an example and a curiosity it can be mentioned that a typical 230 kV breaker tank contains 50 m^3 of oil.

A different oil circuit breaker technology, referred to as minimum oil circuit breakers were installed in large numbers in the 50-ies and 60-ies. They are much smaller and contain much less oil (a few tens of litres per interruption chamber) and as opposed to oil tank breakers, the entire interruption chamber is insulated from ground using solid (porcelain) insulators. The terms *live tank* and *dead tank* are frequently used for these two principally different designs. These terms are general and not limited to oil breakers.

The arcing chamber design in minimum oil circuit breakers is quite different from that of air blast circuit breakers. Figure 4.19 shows schematically the basic design and the principle of operation.

The interruption chamber is a closed cylinder made of a mechanically strong and electrically insulating material, immersed in oil. When the contacts separate, the arc ignites and the oil starts to evaporate and to decompose into hydrogen gas. At first, the gas cannot escape the closed interruption chamber and the pressure rises

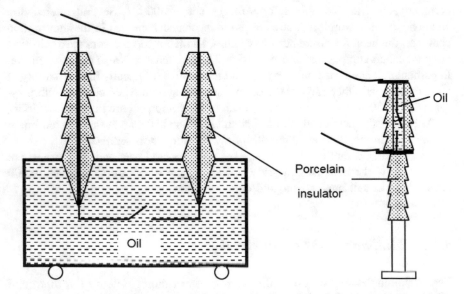

Fig. 4.18 Oil tank (*left*) and minimum oil (*right*) circuit breakers (*schematic*). The driving mechanisms are not included in the drawings

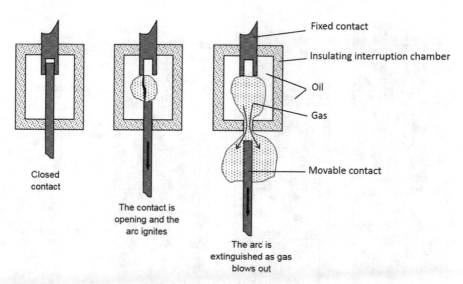

Fig. 4.19 Interruption of current in an oil breaker (*schematic*). The un-shaded areas inside and around the interruption chamber are filled with oil

significantly. As the movable contact is pulled out of the bottom hole, the gas escapes and the high pressure causes a strong blast on the arc, which, therefore, extinguishes. Hence, in this way the arc itself generates the blast that interrupts the current. Due to this, this design is often referred to as the *suicide pot*.

This design is relatively simple, but has its weaknesses. The pressure inside the interruption chamber rises rapidly when large currents are interrupted. If the chamber is not able to withstand the pressure it may rupture before the movable contact is pulled out and thereby relieving the pressure. The interruption then fails and the switchgear most likely explodes. When interrupting small currents, problems caused by too little instead of too high pressure build-up may occur. A too low pressure can result in an insufficient gas blast, unable to extinguish the arc. Again, the interruption is unsuccessful and the switchgear may explode.

Moreover, the interruption capability depends largely on how the current zero crossing coincide with the time at which the moveable contacts is pulled out and opens for the gas blast.

Several of these problems are to a large degree solved by modifying the interruption chamber design compared to the simple and schematic arrangement of Fig. 4.19. Figure 4.20 shows a commercial minimum oil breaker design where a closed volume of gas is compressed, leading to a lower pressure build-up during interruption of large currents. Furthermore, the flow is here in the lateral direction relative to the arc and through several slits, making the interruption ability less dependent on the moment of current zero crossing.

The pressure in the closed gas volume is approximately atmospheric when the breaker is closed. When the arc ignites, the oil pressure rises, leading partly to an oil and gas flow in the gap and partly to a compression of the enclosed gas volume.

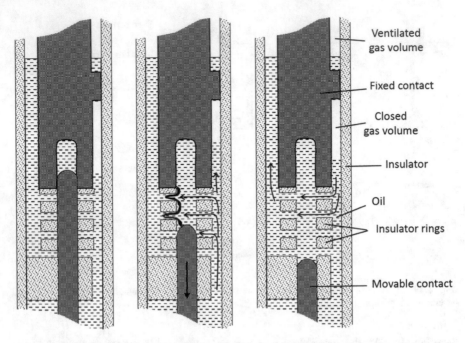

Fig. 4.20 The interruption sequence in a commercial minimum oil circuit breaker; closed (*left*), while the arc is burning (*middle*) and open (*right*)

The arc is blown sideways and into the outlet slits and is eventually, extinguished as the large thermal conductivity of the hydrogen gas causes it to cool down. After the arc is extinguished the gas in the closed volume expands as the pressure in the oil decreases. Clean, fresh oil is thus flushed across the gap between the contacts. Soot particles and other reaction products are in this way removed from the gap, and the dielectric strength is quickly restored.

The ratio between the energy produced by the arc energy and amount of gas formed is almost constant when an arc is burning in oil. Typically, about 60 cm³ gas is formed per kilojoule arc energy. The maximum occurring pressure during an interruption depends on the amount of gas produced and flow resistance and expansion possibilities for gas and/or oil in the interrupting chamber.

To a certain degree, the interrupting capabilities of oil breakers are determined by the mechanical strength of the interruption chamber. It is possible to build chambers that can withstand pressures up to 150 bar by using glass fibre reinforced insulation materials. Driven by material developments, minimum oil breakers for rated voltages of 100–150 kV per interruption chamber have been built [8]. Most minimum oil circuit breakers used for voltages from 145 kV and above have two or more interruption chambers in series As for air blast circuit breakers, grading capacitors in parallel with each interruption chamber are applied.

A clear disadvantage with oil filled circuit breakers is the risk of explosions. Hence, restrictions from governmental bodies on the indoor usage of oil-filled apparatus, including switchgears have been implemented in several countries. The risk of having severe secondary explosions is less if oil filled breakers are installed outdoor.

As previously mentioned, a large number of minimum oil circuit breakers were installed in the period 1965–1980. Their most important advantage, compared to air blast circuit breakers, was the lower price. Even though their maintenance costs are higher than for modern SF_6 circuit breakers, several of these are still in service. Due to few switching operations and robust designs, it may take quite a while before this breaker technology completely disappears.

4.3.4 SF_6 Circuit Breakers

4.3.4.1   Dual Pressure SF_6 Circuit Breakers

The first circuit breakers for transmission level voltages (a few hundred kilovolts) utilizing SF_6 as the interruption medium came around 1960. Their construction and working principle included features from oil or air blast circuit breakers. A valve from a high-pressure gas reservoir opens during opening operations, causing an intense gas flow on the arc as it is pulled through an axial symmetric nozzle arrangement. The gas is collected in a low-pressure reservoir where it is compressed and pumped back into the high-pressure reservoir [9]. The most important distinctions compared to the air blast circuit breaker are that the gas is SF_6 instead of air, and that it is recycled.

The pressure on the high-pressure side is typically 10–20 bar. This can cause problems because parts of the SF_6 may liquefy. This results in a reduction of the gas density and consequently in a decrease of the dielectric strength. For example, SF_6 at 17 bar liquefies already at a temperature of 13 °C, and heating elements are required to keep the breaker operative. Among the other disadvantages are that relatively large amounts of gas are needed, that compressors and high pressure gas handling systems increase the system complexity, leading to more maintenance and higher cost.

Dual pressure SF_6 circuit breakers are also referred to as first generation SF_6 circuit breakers; a large number of them were installed in the 1960-ies in the US and some other countries. In other countries, this circuit breaker technology is nearly completely absent.

4.3.4.2   Single Pressure SF_6 Circuit Breakers

The dual pressure SF_6 circuit breaker was followed by the single pressure SF_6 circuit breaker, which contains no high-pressure reservoir. Except during interruptions, the entire gas volume of the breaker is at the same pressure, typically 4–8 bar.

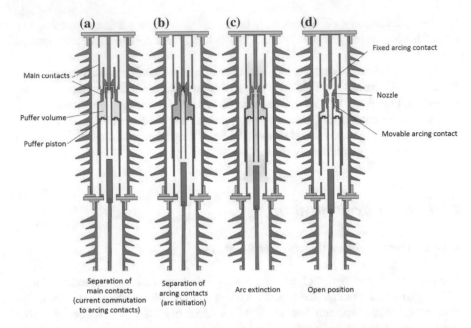

Fig. 4.21 Interruption sequence in a puffer breaker. **a** Separation of the main contacts, **b** separation of the arcing contacts, **c** arc extinction, **d** the breaker in open position

Single pressure SF$_6$ circuit breakers are said to be either of the *puffer* or the *self-blast* type, although the difference in some cases may be somewhat indistinct. In both types, the breaker itself generates the gas flow during the interruption; no external compressors are used. The difference between them lies in what way the energy for the compression and gas flow is provided. Mechanical energy from the driving mechanism is used in the puffer type, while energy from the arc is used in the self-blast type. These breaker technologies are usually termed second and third generation SF$_6$ circuit breakers.

Figure 4.21 shows the interruption sequence of a puffer breaker. The interruption chamber, contacts, nozzles etc. are axial symmetric and placed inside a cylinder filled with SF$_6$.

The main feature of the puffer breaker is the piston/cylinder arrangement, which is an integrated part of the contact system. The piston is fixed, whereas the cylinder moves together with the moving contact. When the breaker is closed, there is a gas volume in this cylinder. When the movable contact is pulled out (down), this volume decreases because the piston remains in its fixed position. This causes a compression of the enclosed gas.

As stated earlier, the puffer breaker normally has two pairs of contacts; one pair that carries current in closed position (*the main contacts*) and one pair that is made of a heat-resistant material which can withstand the thermal stresses and erosion of the arc (*the arcing contacts*). The contact movement is such that the arcing contacts separate last during opening and meet first during closing operation. In this way, the main contacts are not exposed to direct arcing, and the arcing wear is less.

The main contacts, which are placed concentrically outside the arcing contacts, are in Fig. 4.21a open and the current has commutated to the arcing contacts. These are also about to open in Fig. 4.21b and when they do, an arc ignites. A nozzle made of polytetrafluoroethylene (PTFE) or another insulating material is fixed to the moving contact and guides the compressed gas out of the puffer volume and onto the arc just as the arcing contacts separate.

The flow is, to a certain degree, dependent on the amplitude of the current being interrupted. At large short-circuit currents the arc cross-section may be larger than the nozzle throat diameter, blocking the gas flow. This is called *current clogging*. The gas pressure in the puffer volume then continues to increase, mainly because of the mechanical movement, but also due to the transfer of heat from the arc, leading to a rapid temperature rise. As the arc approaches its current zero crossing, the arc cross-section decreases and the compressed gas in the puffer volume flows out of the nozzle creating a powerful blast onto the arc, see Fig. 4.21c.

When interrupting smaller currents the arc cross section is smaller and it does not block the outflow of gas to the same extent. Consequently, the gas pressure does not increase as much and the gas flow is less intense.

After the arc is extinguished, the gas particles recombine, the dielectric strength is quickly restored and the contact movement stops, Fig. 4.21d. The gas pressure then levels out inside the breaker.

The arcing time varies somewhat with the design and is also dependent on how the contact movement coincides with the current zero crossing. The minimum arcing time during interruption with a puffer breaker is about 6–12 ms, while the maximum arcing time is one half cycle longer, i.e. 16–22 ms in a 50 Hz system.

The contact and nozzle design may vary somewhat from one manufacturer to the next. For example, some apply double flow nozzle solutions (see Fig. 4.16) also on puffer breakers.

In addition to moving the lower contact during a switching operation, the driving mechanism in puffer breakers has to provide considerable amounts of energy for gas compression and blowing. SF_6 is a dense and thick gas and the driving mechanism in puffer breakers therefore has to be very powerful. This increases the price and is thus a clear drawback of this circuit breaker technology.

The main motivation for the development of the self-blast (or third generation) SF_6 circuit breaker was the desire to reduce the mechanical power and thereby the cost, of the operating mechanism. The self-blast breakers exploit the heat released from the arc to increase the gas pressure and to generate the gas flow. The working principle of the interruption chamber design in self-blast SF_6 breakers is not very different from the interruption chamber of oil-filled breakers. The arc burns across a contact gap inside a closed chamber and the temperature and pressure greatly increases. When the movable contact is pulled out of the nozzle, the high pressure leads to a strong gas flow onto the arc, which is cooled and extinguished. The flow is therefore generated by temperature increase and the associated pressure rise inside an approximately constant volume, and not as in a puffer breaker where the pressure increase is solely due to a reduction in volume.

As the gas flowing onto the arc is hot, it is difficult to achieve high dielectric strength across the contact gap right after current zero crossing. This is a problem in self-blast circuit breakers. Hot gas has a higher electric conductivity than cold gas and hence the probability of an arc re-ignition becomes greater.

Another disadvantage with this design is that the heating and pressure rise during the interruption of small currents is smaller, and the blowing may be too mild to extinguish the arc. In order to make circuit breakers able to handle both large and small currents, self-blast circuit breaker designs frequently also include a puffer function. An example of such a hybrid design is shown in Fig. 4.22.

When interrupting large currents, the arc is strong and the pressure in the self-blast interruption chamber is therefore high, leading to a powerful blast onto the arc while the contacts separate. The gas, which is compressed in the puffer volume, is released downwards as a valve at the lower end opens.

The pressure build-up in the self-blast chamber is considerably less when interrupting small currents, and also less than in the puffer volume. The valve between the puffer volume and the self-blast volume therefore opens and gas flows up and out through the nozzle. The flow is in this case generated by means of the puffer principle. The valve at the lower end of the puffer volume in this case remains closed.

Whether this breaker design utilizes the self-blast or the puffer principle, depends on the magnitude of the current being interrupted and the resulting pressure

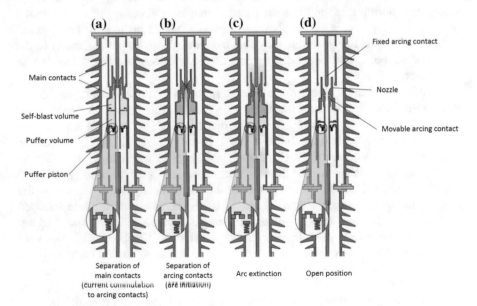

Fig. 4.22 Interruption in a self-blast type SF$_6$ circuit breaker. **a** Separation of main contacts, **b** separation of the arcing contacts, **c** arc extinction, **d** the breaker in open position

build-up. The over-pressure valves in either end of the puffer volume must be carefully balanced for this scheme to work as intended.

When disregarding the arcing chamber design and the power of the operating mechanism, no significant differences exist between second and third generation SF_6 circuit breakers. Both designs function very well, and circuit breakers rated up to 300 kV per interruption chamber are available [10]. For the highest short circuit ratings, the puffer principle is best suited. Some nitrogen is added to the SF_6 gas to lower the point of condensation in switchgear installed outdoor in places where the temperature can reach −30 to −40 °C.

A typical example of a modern SF_6 circuit breaker for 420 kV is shown in Fig. 4.23. It has two interruption chambers in series (one in each of the horizontal porcelain housings) and a separate operating mechanism for each phase. The operating mechanism is located in the cabinet at the bottom of the breaker, together with control and auxiliary circuits. Such circuit breakers are typically rated for continuous load currents of a few thousand amperes, and for interruption of short circuit currents up to 31.5, 40 or 63 kA.

Older designs and designs for the highest ratings (300, 420 kV and above) generally have separate driving mechanisms for each phase (as in Fig. 4.23), while breakers for more modest ratings (145 kV and below) usually have one common driving mechanism for all phases. Shafts and gears then transfer the mechanical

Fig. 4.23 Modern single pressure 420 kV SF_6 circuit breaker (Courtesy of ABB)

power from the driving mechanism and up to the movable contact in each of the three interrupting chambers.

Usually, the circuit breaker manufacturers offer two or three driving mechanism alternatives; hydraulic, pneumatic or spring operated. Spring operated circuit breakers usually have one or more powerful springs that are charged by an electric motor. During a switching operation, the springs are released and the movement is transferred from the driving mechanism to the movable contact by shafts, chains, cranks etc. After operation, the motor runs for some ten seconds to recharge the springs. In hydraulic and pneumatic operated switchgear the energy is released by opening valves in the hydraulic or pneumatic systems, respectively. The mechanical transfer from the diving mechanism to the movable contacts is usually the same.

The driving mechanism is always grounded and the transferral must thus be electrically insulating. This is typically achieved by attaching the moving contact to a long shaft of glass fibre reinforced epoxy, porcelain or another solid insulating material. The driving shaft or rod in the circuit breaker in Fig. 4.23 moves vertically inside the two vertical porcelain insulators during switching operations, and a special gear inside the metal part in the middle of the "T" transfers the movement to the movable contacts in both arcing chambers.

As previously mentioned, the single pressure SF_6 circuit breaker took over more or less the complete market for new installations in the late 1970-ies. The puffer type dominated during the first ten years, but after about 1990, all the major circuit breaker suppliers also offer various self-blast designs. Single pressure SF_6 circuit breakers come in a version for use in gas-insulated substations (GIS) and one version for installation in air-insulated substations, usually outdoors as shown in Fig. 4.23.

Single pressure SF_6 circuit breakers need considerably less maintenance than minimum oil and air blast circuit breakers. Service experience and reliability with these switchgears have in general been very good, although cases exist where puffer breakers have been replaced after only 15 years in service due to lots of flaws and failures. The number of switching operations most of these breakers perform is typically less than a dozen per year, and this causes the wear and tear to be very modest. Moreover, the short-circuit ratings in the transmission grids are not increasing as rapid as earlier, so it is reasonable to assume that many of these circuit breakers will have a service life of at least 40–50 years.

4.4 Distribution Level Switchgear

In medium voltage distribution networks, in contrast to the high voltage transmission networks, as explained in Sect. 4.1, besides circuit breakers, load break switches are used for interruption of load currents, and high voltage fuses fulfil in some cases the function of short circuit current interruption. In the first three parts of this section, load break switches, high voltage fuses and the interaction of these two components are discussed.

Different types of circuit breakers technologies such as minimum oil and SF_6 are also used in medium voltage [11], but vacuum circuit breakers are presently the dominating technology in this voltage range. The last two parts of this section are devoted to different circuit breaker technologies used in medium voltage switchgears, with emphasis on vacuum circuit breakers.

4.4.1 Load Break Switches

Compared to circuit breakers, load break switches are inexpensive, and are extensively used in distribution systems at medium voltage level (typically 6–36 kV); often in series with fuses. Several designs exist, and those dominating today's market either have SF_6 or air as the interrupting medium.

SF_6 load break switches are usually of fairly simple designs, often based on simplified puffer or self-blast principles. The gas pressure is usually only 1–1.3 bar, which is considerably less than that in large circuit breakers. The contact pairs for all three phases are typically located within the same interruption chamber. Gas compression and flow onto the arc is provided by the contact movement and/or heat generated by the arc. No intense gas flow or long insulation distances are needed, as these are switchgear for relatively low ratings. However, it is important that the switchgear design is simple and suitable for inexpensive mass production.

An example of a load break switch with air as interrupting medium is shown in Fig. 4.24. The nozzle is made of a material, which produces gas with a high H_2 content when exposed to an arc. This is referred to as *Hardgas*. As mentioned in relation with oil breakers, hydrogen is a very good interruption medium making the interruption easier than if the arc burns in pure air.

The switchgear has two pairs of contacts, and the driving mechanism is constructed in such a way that the main contacts are the last to mate when closing and the first to separate when opening. The nozzle is narrow causing heat from the arc to generate a certain pressure rise and gas flow as the arcing contact opens. The piston is mechanically linked with the driving rod and generates an air flow onto the arc during interruption.

This and similar breaker designs are very popular in medium voltage distribution systems of medium or low current rating.

The Hardgas principle is only used in load break switches for distribution system voltages. Furthermore, practical breaker designs involving materials, which are supposed to decompose to SF_6, have not reach the market, although patents on this have been granted.

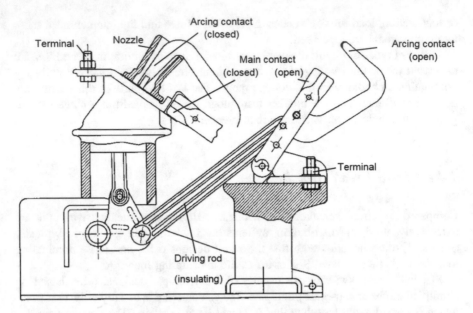

Fig. 4.24 A load break switch with a "Hardgas" nozzle for 6–24 kV; both in closed and open position

4.4.2 High Voltage Fuses

Fuses are inexpensive and provide efficient short circuit protection, as they are able to interrupt very high short circuit currents. Fuses are widely used in medium voltage systems (i.e. up to around 36 kV) with moderate or low load currents. Fuses do not work for higher voltages and/or larger load currents. In such cases, short circuit currents have to be interrupted with circuit breakers. A fuse that has interrupted a current must be manually replaced, and this is an important disadvantage of this technology.

Fuses are reliable for interruption of short circuit currents, but may have difficulties in the over-current range. Over-currents are currents 2–8 times greater than the nominal load currents and consequently much smaller than short circuit currents. In such cases, the fuse element, typically a silver wire, *melts* and an arc is ignited, but the device may not be able to *interrupt* the current.

There are two main types of fuses: *current-limiting* and *non-current-limiting*. Non-current-limiting fuses interrupt the current at current zero crossing, just as in most switchgears. The interruption medium may be a gas or a fluid.

In current-limiting fuses on the other hand, a large voltage builds up across the fuse and causes the current to be limited and forced to zero before the natural current zero crossing. The short circuit current does not reach its prospective peak value, giving the advantage of reduced mechanical stresses as the electromagnetic forces generated by the current become lower.

Fig. 4.25 Current-limiting
high voltage fuses (Courtesy
of ABB)

A distinction is made between *general purpose* and *back-up* fuses. General-purpose fuses shall interrupt all currents, which cause melting of the fuse wire. The manufacturer of back-up fuses must state a minimum breaking current. In medium voltage distribution systems, current-limiting back-up fuses are mainly used and these will be the subject of this section.

A photo of some fuses is shown in Fig. 4.25. The terminals are made of metal to provide good electrical contact to the fuse holder (not in the photo). The fuse body or housing itself is made of porcelain.

The current ratings are in international standards recommended to be limited to a load current of 400 A, but most manufacturers only provide fuses for currents up to 100–200 A for high voltage applications. To obtain higher rated currents several fuses may be connected in parallel.

4.4.2.1 Design and Principles of Operation

A fuse is in many respects a component of a very simple design (see Fig. 4.26).

In high voltage fuses, it is often necessary to have fuse elements that are longer than the fuse body. This is realized by winding the fuse elements on an electric insulating core. This winding core has a star-shaped cross-section to make sure that the fuse element is surrounded by sand more or less on all sides. The fuse element is made of silver, and commonly several elements are put in parallel. The cost of the silver can amount to as much as half the production cost for a high voltage fuse, but a less expensive material that can replace silver has not yet been found. It is

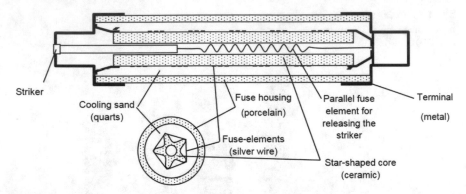

Fig. 4.26 Principle drawing of a high voltage fuse. A lengthwise section (*upper*) and cross-section (*lower*)

particularly the good properties of silver at high temperatures (little corrosion/oxidization) and the good electric conductivity which make it very suitable as a fuse element.

The sand should be quarts (SiO_2), and the quality with regard to purity, grain size and particle shape is of great importance for its arc cooling properties.

Fuses are often equipped with an alarm system for indication and for initiation of a breaker operation. This may be achieved by placing a relatively poorly conducting wire in parallel with the fuse element. As the fuse element melts, the current commutes to the parallel wire inside the winding core. This then quickly melts off and releases the spring charged striker pin. The striker pops out and can be used to automatically trip a switchgear.

In a current-limiting fuse, a large number of arcs connected in series are formed, and the total arc voltage effectively forces the current to zero. In the past, the fuse elements were wires or band with constant thickness or widths. At short circuit the entire elements thus melt instantaneously and surface tension and magnetic pressure cause the melted wire to split into many droplets, each connected by a short arc. Consequently, a large number of arcs in series were produced and the current was quickly forced to zero. However, because the *di/dt* then becomes very large, considerable over-voltages were generated, and this is an obvious disadvantage. Fuse elements of constant cross sectional area are also a problem as the fuse becomes very warm at nominal load current.

Modern fuses are therefore made with fuse elements that are in the form of bands with evenly distributed narrow sections. The melting at short circuit currents occurs at these narrowings, and a predetermined number of arcs in series are obtained. Consequently, a better control over the arc voltage and also of the over-voltages occurring when interrupting short circuits currents is achieved. It has been found that about 5–8 arcs per kilovolt rated voltage is adequate. The arc is cooled by the quarts sand giving an arc voltage drop of about 200–300 V per arc.

At nominal load current and lasting over-currents, the high thermal conductivity of silver may lead to a decreased temperature difference between the narrowings and the rest of the fuse element. Examples of fuse element designs are shown in Fig. 4.27. The local temperature distributions at short circuit and over-currents ($3 \times I_N$) at the time of melting are also shown. Silver melts at 960 °C.

Heat conductance is insignificant at short circuit so most of the heat is found where it is generated (see Fig. 4.27). In these cases, all the narrowings melt within a very short time period.

When considering the *entire* fuse element at persistent over-current conditions, on the other hand, heat conductance from the fuse to the surroundings is important for the temperature distribution. Figure 4.28 shows the temperature variation along the entire length of the fuse element at over-currents.

Heat is carried away through the fuse terminals. The temperature in this area is thus somewhat lower than that in the middle of the fuse. Consequently, melting at over-currents first occurs in the middle region, possible only a few arcs are formed, and they may combine into one arc that is gradually elongated. Meanwhile, the sand is also heated and starts melting. There is first an increase in the arc voltage due to the increasing arc length, followed by a decreased arc voltage due to poorer cooling efficiency. However, the arc voltage is much smaller than when an arc forms in every narrowing. The current interruption may therefore fail, and the arc may eventually burn through of the fuse terminals. As a result, flashover between the phases and a full short circuit in the system may follow.

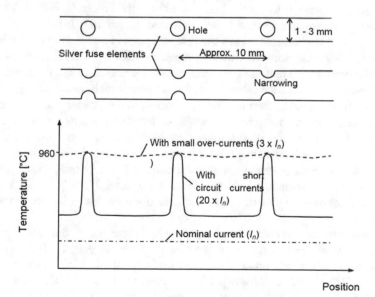

Fig. 4.27 Examples of fuse element geometries (*upper figure*) and their temperature profiles (*lower figure*) as they melt at short circuit currents (*solid*) and over-currents (*upper dashed line*)

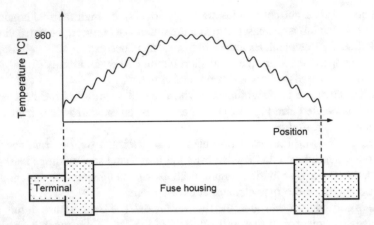

Fig. 4.28 Temperature in the fuse element in the different parts of the fuse at the moment of the melting (*schematic*)

Obviously, it is desirable to have the lowest possible ratio between the minimum breaking current and the nominal rated current of a fuse. That is, the fuse should preferably interrupt not only short circuit currents, but also over-currents. The minimum breaking current may be reduced by connecting several fuse elements in parallel. Assume, for example, that a fuse element rated for nominal load current of 25 A melts at 50 A, but needs at least 200 A of current for a successful interruption. The minimum breaking to nominal current ratio is then eight ($I_b = 8 \times I_N$).

If four of these fuse elements are connected in parallel, a fuse with 100 A nominal current is obtained. One of the fuse elements is assumed to melt when the fuse carries twice the nominal current, i.e. approx. 200 A. This fuse element will then cease carrying current, and the current is commutated to the three remaining fuse elements. One after the other, these fuse elements melt, since the current they carry exceeds the melting current. The last element to melt carries the entire current of 200 A alone, and is therefore also capable of *interrupting* the current. This could lead to re-ignition in one of the other fuse elements and the current may thus commutate from one fuse element to the next a few times (re-igniting the arcs) until the dielectric strength is great enough for a definite interruption. In such a way, a fuse with 200 A minimum breaking current is obtained. ($I_b = 2 \times I_N$.) This is a considerable improvement from what is obtained with a fuse with one single fuse element.

At over-currents, the entire fuse becomes very warm before the hottest point of the fuse element reaches the melting point of silver at 960 °C. However, it is possible to influence this process by attaching a small piece of tin onto the fuse element [12], a so called "*M-spot*". The tin melts at about 250 °C, but already at lower temperatures it diffuses relatively quickly into the silver and intermetallic compounds, particularly Ag_3Sn are formed. These compounds (alloys) have a relatively high resistivity, leading to a local resistance increase of the fuse element, which then melts after some time at about 250 °C.

Consequently, fuses equipped with M-spots generally have lower maximum temperatures and lower losses than fuses without M-spots. The losses at nominal current in a regular 100 A/12 kV fuse are about 250 W. This is reduced to less than half in an equivalent fuse with M-spots.

4.4.2.2 Fuse Characteristics

The *time/current characteristic* gives the fuse interruption time as a function of the prospective current, i.e. the current that would flow in the circuit if it were not being limited by the fuse. The time/current characteristic is, for large currents, closely coinciding with the fuse *melting characteristic*. In other words, at large currents the fuse interrupts just after the fuse elements melt. In the over-current range, the deviation between melting and breaking times is greater, and the characteristics thus become different in this region.

The melting characteristics of high voltage fuses of nominal current in the 6–200 A range from one supplier are shown in Fig. 4.29. The characteristics are dashed in the region where interruption cannot be guaranteed, i.e. at small over-currents.

The time/current characteristic may vary considerably from one manufacturer to the next. This must be taken into consideration when coordinating with other protective device, e.g. over-current relays, and when assessing the risk of blowing fuses under transformer inrush currents.

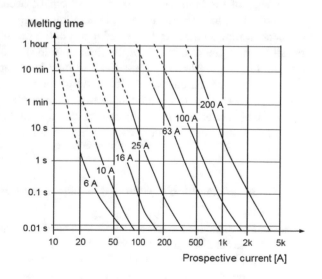

Fig. 4.29 An example of melting characteristics for "back-up" high voltage fuses. The time is calculated from cold state (20 °C). The nominal or rated load currents of the fuses are given in the figure

Fig. 4.30 Example of corresponding voltage (*upper*) and current (*lower*) waveforms when a fuse interrupts. The fault current is here strongly asymmetric. The time t_1 is when the fuse element melts; t_2 is when the arc voltage has the same magnitude as the system voltage, t_3 is when the arc voltage reaches its peak value and the current is interrupted at t_4

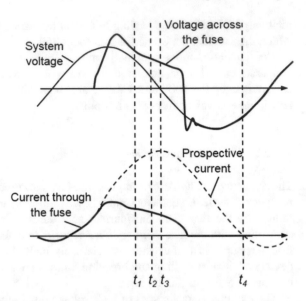

Figure 4.30 shows typical current and voltage waveforms when a fuse interrupts a current. When the total arc voltage across the fuse is of comparable magnitude as the system voltage, the current amplitude during arcing becomes smaller than the prospective current.

Due to inductance in the circuit, the voltage across the fuse may be greater than the system voltage. At short circuit, the time until melting depends on the degree of asymmetry of the fault, i.e., when the short circuit occurs.

As opposed to most switchgears, fuses limit the fault currents considerably. An efficient current limiting leads to a reduction in the mechanical and thermal stresses on all the equipment that carries the fault current. For a symmetrical short circuit current, the *current-limiting characteristic* may be determined precisely, showing the peak value of the current as a function of the prospective short circuit current. Such a characteristic for the product line of one manufacturer is shown in Fig. 4.31.

As the characteristic shows, fuses with low nominal load currents give a greater current limitation effect than fuses with higher nominal currents.

4.4.3   *Interaction Between Fuse and Load Break Switch*

A general purpose load break switch should be able to interrupt load currents, small inductive currents (no-load transformer), small capacitive currents (no-load lines, cables etc.). Moreover, it should be able to close against a short circuit in the system and carry the short circuit current for one to three seconds.

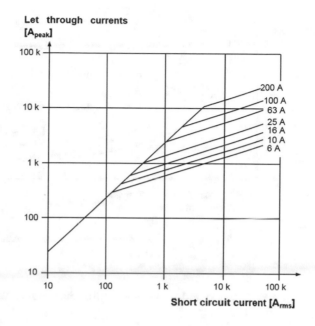

Fig. 4.31 Current limiting characteristic for a certain series of current-limiting high voltage fuses. The nominal load currents are given in the figure

Such a switching component in series with a fuse is very commonly used in medium voltage distribution systems. Load currents are interrupted with the load break switch (and the fuse being totally inactive and unaffected), and short circuit currents are interrupted with the fuse. In the case of over-currents, the fuse elements melt and thereby release the striker, which in turn trips the load break switch.

At low over-currents, the fuse element melts approximately at the middle of the fuse where the temperature is greatest, see Fig. 4.28. As previously described, this may result in only one arc and the fuse may not be able to interrupt. However in any case, the striker trips the load break switch operating mechanism, and its contacts start opening and it will eventually interrupt the current. The operational time for the striker and load break switch combination is around 50–80 ms, meaning over-currents are interrupted by the load break switch well before the arc burns through the terminals of the fuse. This takes more than 200 ms.

Under short circuit currents, the fuse quickly and effectively interrupts. The striker again trips but in this case, the load break switch acts only as a disconnector switch (because the current has already been interrupted by the fuse).

The operation modes of the fuse/load break switch combination thus depend on the magnitude of the current. For large currents the fuse interrupts (the fuse breaking time = the fuse melting time + the fuse arcing time). For small currents the breaker interrupts (the breaker interruption time = the fuse melting time + the striker operation time + the breaker operation time).

Consequently, if the interruption time for the fuse is shorter than for the load break switch, the fuse interrupts and vice versa. At a certain current, the so-called *transfer current*, the interruption times for the fuse and breaker are the same. Thus,

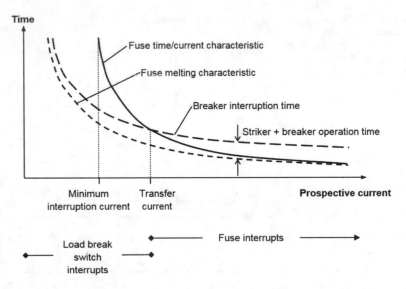

Fig. 4.32 Operating modes for a fuse in series with a load break switch (*schematic*)

the breaker must be able to interrupt currents up to the transfer current, while the fuse must be able to interrupt all currents larger than this.

The various fuse/load break switch interactions are illustrated in Fig. 4.32.

The interruption time for the breaker is closely correlated to the fuse melting characteristic. The fuse time/current characteristic is for high currents coincident with the melting characteristic. As the minimum interruption time for the fuse is approached, the arcing time increases. The transfer current is always higher than the minimum interrupting current, meaning interruption at the entire current range is guaranteed.

4.4.4   Circuit Breakers with SF₆, Oil and Air as the Interrupting Medium

Before SF_6 entered the market, minimum oil circuit breakers were a very competitive technology also for medium voltage applications, i.e., distribution systems and industrial networks. The design and functional principles are largely the same as for minimum oil breakers for higher voltages, except that only one interruption chamber is required.

With regard to SF_6 circuit breakers for medium voltages, their design includes in many cases features from the puffer and self-blast circuit breakers for higher voltages, and in many cases they are considerably different [13]. An example of a design where the gas cooling of the arc is due to movement of the arc itself (and not the gas blowing) is shown in Fig. 4.33.

Fig. 4.33 The contacts in a SF$_6$ circuit breaker, which can interrupt 25 kA at 12 kV

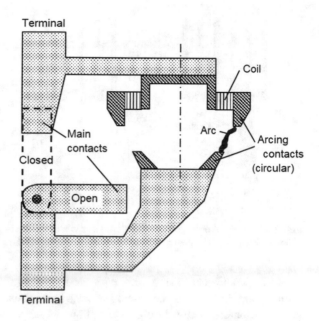

When the main contacts open current commutates to the arcing contacts, and by separation of the arcing contacts, an arc ignites. Before the current reaches the arcing contacts, it passes through a coil. A magnetic field B directed across the contact gap is established due to the current in the coil. The current in the arc is associated with an electric current density J across the contact gap. The magnetic field and the current interact and form an electromagnetic force ($F = J \times B$) on the arc, pulling it in a tangential direction. Hence, this causes the arc rotate on the circular arcing contacts, and the desired cooling is achieved.

The advantages of this design are the modest requirements on the driving mechanism, and that the rotation of the arc reduces the arc erosion on the contacts, so the lifetime of the device becomes long.

There are also circuit breakers for the medium voltage level utilizing air as the interruption medium. These are often based on increasing the arc voltage by splitting the arc into several shorter arcs and thus increasing the number of anode and cathode voltage drops (see Fig. 2.9). The arc voltage can also be increased by designing the interruption chamber in such a way that electromagnetic forces or gas flow drive the arc into contact with metallic or ceramic plates or other parts designed to take up heat from the arc and thereby cool it. Switchgears in which the arc is moved by electromagnetic forces are often referred to as *magnetic air circuit breakers*.

Figure 4.34 shows a schematic view of a medium voltage breaker which combines the two effects just described (splitting the arc and forced cooling against heat resistant plates). The magnetic coils and the bellows that make the arc move are not included in the drawing.

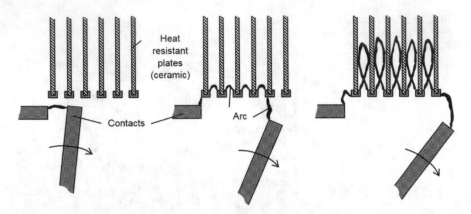

Fig. 4.34 Opening operation in a circuit breaker where the arc is split and cooled against ceramic plates (*schematic*). The device which forces the arc towards the plates is not drawn

Such switchgears have a very high arcing voltage. They thus become current limiting, and in addition, the circuit becomes more resistive. The disadvantage is the large dissipated energy during interruption, which causes the breaker to become relatively large.

Magnetic air switchgears are still in use in some applications with high load currents and frequent switching operations, such as industrial systems. They are less common in ordinary distribution systems, and in recent years, this technology has lost market shares to SF_6 and vacuum interrupters.

The same technology is also used for low voltage switchgear, where an arc in atmospheric air is ignited by the contact separation and pushed with a magnetic force towards a number of insulating or metallic plates, which are referred to as *arc chute*, to enhance the cooling by arc elongation or to increase the arcing voltage by arc splitting. This is a common low voltage switching technology and finds application in a wide range of switches with ratings from few amperes to ten (or even hundreds of) thousands of amperes.

4.4.5 Vacuum Interrupters

4.4.5.1 Introduction

In the breakers which have been described till now, pressurised gas is used as cooling medium and a high gas density is used to increase the dielectric strength in the contact gap after the arc has been extinguished. Very similar features can be achieved by using gas at extremely low pressures. In a vacuum of about 10^{-6} Torr (equal to 10^{-9} bar) the particles in the arc quickly and unhindered move out of the contact gap making the thermal time constant small. The absence of particles in the contact gap also leads to a great dielectric strength immediately after the arc is extinguished. Switchgears using vacuum in the interrupting chamber have become very popular, but only for voltages up to about 52 kV. This is mainly because much larger gap lengths are necessary for higher voltage ratings, as described in Sect. 4.2.5.

As a curiosity, it can be mentioned that the first patent granted for a vacuum interrupter was based on the idea that as long as there is no gas present there could not be an arc established when the contacts separated and interruption would consequently become very straightforward. Later, more thorough investigations showed that an arc is formed in vacuum as well, and that it burns in a metal vapour from the arc foot points at the contacts.

4.4.5.2   Vacuum Interrupter Design

A principle drawing of a typical vacuum interrupter is shown in Fig. 4.35. The spring arrangement, which provides the contact movement, i.e. the driving or operating mechanism, is not included. The interrupting chamber or *vacuum bottle* is a 10–30 cm tall, axial symmetric device with conductor terminations at the top and bottom.

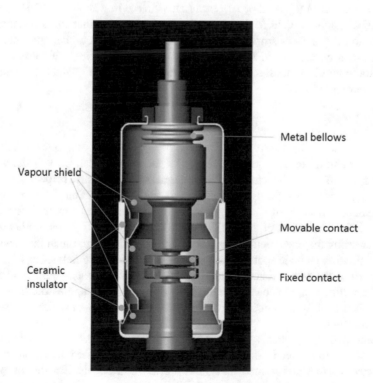

Fig. 4.35   Cut-away drawing of a vacuum interrupter in closed position (Courtesy of ABB)

The design of a vacuum interrupter is very simple. There are only one pair of contacts responsible for both current carrying in the closed position and current interruption. Two terminals of the vacuum interrupter are insulated from each other using one or two insulators made of porcelain or glass. The metal bellows near the flange facilitate the necessary 1–2 cm movement of the upper contact without destroying the vacuum. To avoid covering of the inner surfaces of the insulator by the condensation of highly conductive metal vapour, one or more metallic bodies are used to shield the insulator, which are referred to as *vapour shields*. The chamber is degassed, evacuated and sealed for life when delivered from the factory. Hence, the very good vacuum is assumed to last for the entire service life. This puts strict requirements on gaskets and seals, and in particular, on tight joint between porcelain and metal parts. Moreover, it is necessary that the electrode materials only contain very minute amounts of contamination elements that over time can come out as gas.

Experiments have shown the magnitude of the current where the transition from diffuse to constricted arc occurs can be raised considerably if there is an axial magnetic field in the contact gap. As mentioned in Sect. 2.4, a diffuse arc is much easier to handle than a constricted arc, so this is an important issue when designing vacuum interrupters. The most common method to generate an axial magnetic field is to make slits in the contact members in such a way that the current path to a certain extent takes a helical or tangential shape, like in a solenoid. This gives rise to an axial magnetic field in the contact gap, resulting in that the unwanted transition from diffuse to constricted arc may first occur at much higher currents. In this case, the charge carriers in the gap are exposed to two types of forces, namely electrostatic and Lorentz forces. The equation of motion of charged particles may be then expressed as:

$$m\frac{d\mathbf{v}}{dt} = q \cdot (\mathbf{E} + \mathbf{v} \times \mathbf{B}) \tag{4.4}$$

Here, m and q are the mass and charge of the charged carrier, $\mathbf{v}$ is the velocity vector, E and B are the electric field and magnetic flux density, respectively. In the case of axial magnetic fields, both E and B are in z-direction. This results in a helical shape trajectory of charged particles, e.g. electrons, as shown schematically in Fig. 4.36a. The radius of this helical trajectory known as *Larmor radius* reduces with increasing magnetic field. Thus, the charged particles are much more bound to their originating cathode spots in case of high axial magnetic fields, see Fig. 4.36a, and therefore the interactions between adjacent cathode spots resulting in a transition from diffuse arc to constricted arc is to a great extent suppressed. In this way, it is possible to shift the transition from the diffuse arc to the constricted arc to much higher currents.

An alternative approach is to accept the presence of a constricted arc, but instead take measures to reduce its damaging effects. It turns out that interactions between the electric current and its magnetic field can be utilized also for this purpose. If there is a radial magnetic field in the gap between the contacts, it will be

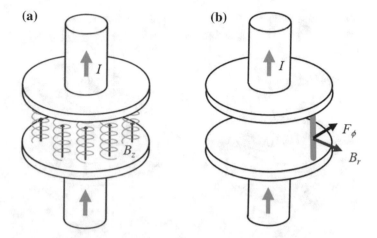

Fig. 4.36 Schematics of vacuum arc—magnetic field interaction in case of **a** axial magnetic fields and **b** radial magnetic fields

perpendicular to the arc. Therefore, Lorentz forces ($F = J \times B$) generate a tangential electromagnetic force on the arc, leading to a rapid rotation of the arc on the contact, see Fig. 4.36b. This reduces the melting of the cathode surface since the anode spot is continuously moving.

A radial magnetic field in the gap may be generated by modifying the design of the contact members. If the current, when flowing through the electrodes, by use of slits is forced to run in the tangential direction along the outer edge of one of the contact members and in the opposite tangential direction in the other member, this sets up a radial magnetic field component in the gap. Examples of how the electrodes can be modified to achieve radial magnetic fields in the gap between the contacts are shown in Fig. 4.37. These design principles are very commonly applied in vacuum interrupters [14, 15].

An axial magnetic field in vacuum interrupters is either generated by modifying the current path [16–18], as by radial magnetic field contacts, or by using an external coil which is place in series to the vacuum interrupter contacts [19].

Strict requirements are put on the contact material in vacuum interrupters. The material must be completely degassed during production as later emissions of gas molecules from the metal can destroy the vacuum. No oxide layer is formed on the contact surfaces in vacuum, causing the contact members to easily weld together during contact mating. Hence, non-welding materials must be used. It is also important that the contact resistance is low when the breaker is closed. Heat generated in the contact interface has to be carried away by thermal conduction through the current leads. Consequently, removing heat from the contacts in vacuum is a greater challenge than when the contacts are surrounded by oil or gas as in other types of switchgear.

In addition, the contact material has a significant impact on the current interruption capability of the vacuum circuit breaker. When the current decreases

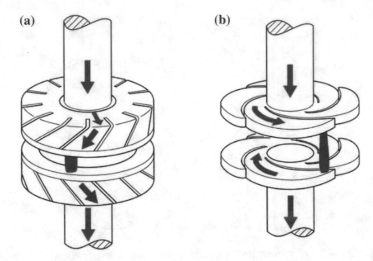

Fig. 4.37 Examples of contact members in vacuum interrupters with radial magnetic fields and the corresponding current paths **a** cup contacts and **b** spiral contacts

towards the end of the half cycle, the arc returns to its more harmless diffuse mode. The cathode spots die one by one, and finally, there is only one spot left. This last spot abruptly ceases to exist just before the natural current zero crossing, causing current chopping. Current chopping is a typical vacuum interrupter phenomenon, although it can occur under certain conditions in other types of switchgears as well. The chopping current in a vacuum interrupter is determined by the contact material and varies within the range of 2–15 A [20]. Current chopping is an unwanted phenomenon since such a rapid change in current may lead to overvoltages. A low chopping current is therefore favourable when choosing contact materials for vacuum interrupters.

The most common contact materials in vacuum contactors, which are used to interrupt load currents, are copper with a few per cent bismuth. Adding small amounts of bismuth radically changes the welding properties of copper and prevents contact welding during closing. In vacuum circuit breakers, a mixture of copper and chrome is used as contact material. The proportions differ from manufacturer to manufacturer, ranging from 50 to 75 wt.% of copper.

4.4.5.3   Applications of Vacuum Interrupters

Vacuum interrupters have become very popular, mainly as circuit breakers for voltages between 6 and 36 kV, and are available for load currents up to 5000 A and for interrupting short-circuit currents up to 63 kA. In industry and distribution systems where the ratings are too large for applying a load break switch/fuse

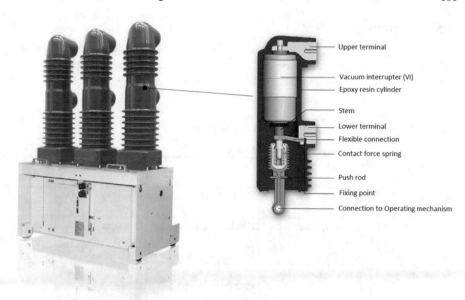

Upper terminal

Vacuum interrupter (VI)
Epoxy resin cylinder

Stem

Lower terminal
Flexible connection

Contact force spring

Push rod

Fixing point

Connection to Operating mechanism

Fig. 4.38 Vacuum switchgear using embedded poles of vacuum interrupters (Courtesy of ABB)

combination, vacuum interrupters have nearly the entire market for new installations.

For higher voltages and/or currents, e.g. generator circuit breakers, it is possible to put several vacuum interrupter units ("bottles") in series and parallel. Circuit breakers for the 145 kV level using two vacuum interrupters in series also exist, but are common only in a few regions of the world. For both these applications SF_6 based technology offers a strong competition.

The main advantages with vacuum switchgears are that they are reliable, very compact, almost maintenance free and have a very long service life (tens of thousands of interruptions). Their disadvantages are primary that they are somewhat expensive and that special precautions (surge arrestors) may be needed to handle over-voltages caused by current chopping and multiple re-ignitions.

A typical commercial vacuum switchgear is shown in Fig. 4.38. The vacuum bottles are mounted in a so-called *embedded pole*. Here, epoxy resin is used to cover the outer surface of a vacuum bottle and in this way to enhance its dielectric and mechanical characteristics. The moving electrode of the vacuum interrupter is connected through a flexible connection to the lower terminal of the pole. The mechanical motion is transferred through an insulating push rod from the drive mechanism to the moving contact. Other manufacturers use vacuum bottles in vacuum circuit breakers without embedding them. In those designs, the vacuum bottles are mechanically fixed to a solid insulating structure.

4.5   Summary

Tables 4.1 and 4.2 summarise the most important switchgear technologies, together with a list of their most important advantages and disadvantages.

Table 4.1 Circuit breakers for the highest voltages (U > 70 kV)

	Single pressure SF$_6$	Minimum oil	Air blast
Insulating medium	SF$_6$ at 5–8 bar	Oil	Air typically at 20 bar
Interrupting medium	SF$_6$	H$_2$, H, C$_2$H$_2$, etc.	N$_2$
Arc cooling principle	Gas flow (puffer and self-blast)	Gas flow, oil flushing	Gas flow
Arc voltage	Low (<1 kV)	Very high (many kV)	High (several kV)
Disadvantages	SF$_6$ is a "greenhouse gas" Toxic dissociation products	Many interruption chambers in series Need for a mechanically strong chamber. Much maintenance. Fire hazard	Many interruption chambers in series Noise when switching. Much maintenance
Advantages	Up to about 300 kV per interruption chamber. Little maintenance. Very reliable	Inexpensive	Compressed air both as insulating and interrupting medium, and in the driving mechanism

Table 4.2 Some common switchgears for medium voltages (6 kV < U < 70 kV)

	Vacuum	SF$_6$ self-blast or rotating arc	Minimum oil	SF$_6$	Switchgear with "hard-gas"
Application	Circuit breaker	Circuit breaker	Circuit breaker	Load break switch	Load break switch
Insulating medium	Vacuum	SF$_6$	Oil	SF$_6$ at 1–2 bar	Air
Interrupting medium	Metal vapour	SF$_6$	H$_2$, H, C$_2$H$_2$, etc.	SF$_6$	H$_2$ and air
Arc cooling principle	Movement of the arc due to magnetic forces	Gas flow, movement of the arc due to magnetic forces. Arc splitting	Gas flow, oil flushing	Gas flow	Dissociation of gas. Gas flow
Arc voltage	Very low	Low	High	Low	High

Exercises

Problem 1

At what ratings (I_n and U_n) are fuses used in power systems? What are the main advantages and disadvantages by using fuses instead of breakers to interrupt current?

Sketch how a typical high voltage fuse is designed, indicate what materials are used and briefly explain the functions of the main parts.

Problem 2

The circuit diagram of Fig. 4.39 will be used to analyse the interruption of a short circuit current with a fuse in a 50 Hz single-phase network. The short circuit is modelled by a switch in parallel to the load that is closed at the moment of the short circuit.

For the entire exercise assume that the short circuit happens at the source voltage zero crossing ($t = 0$). The short circuit current is much larger than the load current so sufficient accuracy in the calculations is achieved by setting $i(0) = 0$.

Find an expression for the prospective fault current (the current that goes through the fuse assuming it is being replaces with a lossless conductor).

How large becomes the prospective current's maximum value when $U = 12$ kV and $L = 6.37$ mH?

Problem 3

Assume that the fuse element warms up adiabatically from an initial temperature T_0. Assume that the heat capacitance c and the resistivity ρ of the fuse element are constant (temperature independent) during the heating process. Derive an expression for the fuse melting time t_s as a function of network parameters, fuse element cross-section A, as well as its temperature and material proprieties.

Hint: Sufficient accuracy is achieved by assuming that $\cos x = 1 - x^2/2$ for small values of x.

Assume an initial temperature $T_0 = 60\,°C$, and insert the given values of the network parameters U, L, and ω, together with the following material parameters:

resistivity $\quad\quad\quad \rho = 3 \times 10^{-8}\ \Omega\ m$
density $\quad\quad\quad\quad \eta = 10{,}500\ kg/m^3$
heat capacitance $\quad c = 235\ J/kg\ K$
melting heat $\quad\quad c_m = 105{,}000\ J/kg$

Fig. 4.39 Figure of problem 2

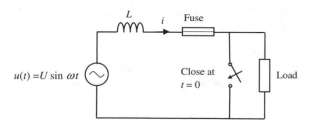

melting point $T_m = 960\ °C$

and show that with $A = 0.7\ \text{mm}^2$ the fuse element melts 5 ms after the short circuit occurred.

Problem 4
When the fuse element melts after 5 ms a number of arcs are ignited in the fuse. The arcs cause a voltage drop across the fuse, u_b. Find an expression for the current through the fuse while the arcs are burning (assume u_b constant).

How large must u_b be to reduce the interruption time to 1.5 ms after the melting of the fuse element?

Hint: Enough accuracy is achieved by assuming $\sin x = x$ *for small x.*

Sketch the current waveform from the instant of the short circuit to the instant of the current interruption.

Problem 5
Often the melting element has distributed narrow sections. What is the function of these narrowings?

It is important to adjust the number of narrowings to the rated voltage of the fuse. Explain briefly what problems can arise if the number of narrowings is too high at a given rated voltage.

It is also usual to attach a small piece of tin to the melting element (so called "M-spot"). Explain briefly their function.

Problem 6
A popular breaker solution in 12 kV grids is a load break switch in series with a fuse equipped with a striker. Explain how the breaker and fuse work together during interruption of:

(i)   Load currents,
(ii) Moderate over currents (like $2I_n$ where I_n is maximum load current in the circuit),
(iii) Short circuit currents.

In which of these cases will it be necessary to replace the fuse afterwards?

Problem 7
Another popular switchgear technology for 12 kV grids is the vacuum interrupter. Explain briefly and by use of a figure how a vacuum interrupter is constructed.

The electric arc in a vacuum interrupter can assume two different modes or states. What are these called, and which parameter is decisive for what mode the arc takes. What characterises the appearance of the arc and the voltage drop in the two cases?

Problem 8
SF$_6$ circuit breakers are commonly sorted into three generations. Describe briefly the characteristics of each generation with regard to gas blow/arc extinction

principle, and do also outline the major advantages and disadvantages associated with each generation.

This breaker employs, like many breaker designs do, two sets/pairs of contacts. What is the reason for using two contact sets, and what are they usually called?

What is the main technical requirement for each of the two contact sets? Give examples of materials or material combinations that are being used in the two cases.

In what sequence do the two contact sets mate during closing and separate during opening operations?

Problem 9

Oil is a common arc quenching medium. Describe briefly—for example by including a schematic drawing showing the various substances in and nearby the arc—the conditions observed after a pair of current carrying breaker contacts immersed in oil is separated.

Why is it so important to keep the arc immersed in oil and thereby avoid contact with air?

During arcing in oil soot or carbon particles are produced. Why is this a clear disadvantage with this quenching medium?

Problem 10

Describe briefly and by means of a drawing how the arcing chamber of a vacuum interrupter (the "vacuum bottle") is built up.

What are the main areas of application (current and voltage ratings, types of power networks) of vacuum interrupters? What are their main advantages compared to competing technologies?

Problem 11

Substations in the Norwegian 145, 300 and 420 kV systems are a mixture of gas insulated substations (GIS) and air insulted substations (AIS). What is the single most important reason for choosing GIS instead of AIS technology, and at what kind of locations are these GISs typically installed?

What are the most important drawbacks with the GIS compared to the AIS technology.

The solid insulation inside GIS is provided by epoxy supports. Why are these normally bell-shaped? Besides providing dielectric insulation they serve a second purpose. Which purpose?

References

1. MacDonald J (ed) (2012) Electric power substations engineering, 3rd edn. CRC Press, Boca Rayton
2. IEEE Standard 605-2008 (2008) IEEE guide for bus design in air insulated substations
3. Lee Willis H (2004) Power distribution planning reference book, 2nd edn. Marcel Dekker Inc., New York, Basel

4. ICF Consulting (2002) By-products of sulphur hexafluoride (SF_6) use in the electric power industry
5. Latham R (ed) (1995) High voltage vacuum insulation: basic concepts and technological practice. Academic press, London, San Diego, New York
6. Kesaev IG (1959) Sov Phys Tech Phys 4:1351
7. Rankin W, Bennett R (1941) A conserved-pressure air-blast circuit breaker for high-voltage service. Electr Eng 60:193–196
8. Gugelmann U (1968 New developments in high voltage minimum oil circuit breakers. IEEE Trans Power Apparatus Syst PAS-87:1613–1622
9. Yeckley R, Cromer C (1970) New SF_6 EHV circuit breakers for 550 KV and 765 KV. IEEE Trans Power Apparatus Syst PAS-89:2015–2023
10. ABB Live tank circuit breakers, application guide, 2014
11. Stewart S (2008) Distribution switchgear, IET power and energy series 46. London, UK
12. Wright A, Newbery P (2004) Electric fuses, 3rd edn. IET power and energy series 49, London, UK
13. Pary J (1986) Development of SF_6 switchgear incorporating rotating-arc circuit breakers. Electron Power 32:602–604
14. Schneider H (1960) Contact structure for an electric circuit interrupter, US patent 2,949,520
15. Smith S (1963) Contact structure for an electric circuit interrupter, US patent 3,089,936
16. Mayo S (1998) Axial magnetic field coil for vacuum interrupter, US patent 5,777,287
17. Schellekens H, Shang W, Lenstra K (1993) Vacuum interrupter design based on arc magnetic field interaction for horse-shoe electrodes. IEEE Trans Plasma Sci 21:469–473
18. Homma M et al (1999) Physical and theoretical aspects of a new vacuum arc control technology-self arc diffusion by electrode: SADE. IEEE Trans Plasma Sci 27:961–968
19. Sato T et al (1995) Vacuum circuit breaker, US patent 5,379,014
20. Slade P (2008) The vacuum interrupter: Theory, design, and applications. CRC Press, Boca Rayton

Chapter 5
Service Experience and Diagnostic Testing of Power Switching Devices

The previous chapters looked into various aspects of current interruption in power systems, such as different switching duties and the associated recovery voltages, properties of the electric arc, and design and technologies applied in switching devices. The present chapter is concerned with the service life of the switchgear. More precisely, it elaborates on what electric utilities and industry can do to ensure that the breakers installed in their power network serve them well during the devices' many years of service after installation and commissioning. Carrying out the appropriate maintenance and in other ways managing these assets in a good and cost-efficient way is an important task.

In general, the power switching devices in service or available on the market are examples of mature technologies. Their design relies on well-proven principles and they are made of carefully selected sub-components and materials. Moreover, they are manufactured, assembled, erected and put into service under a strict quality assurance regime. Hence, most switchgear in service today are very reliable, require little maintenance and have an expected lifetime of several decades and thousands or even tens of thousands of switching operations. This is particularly the case for the circuit breakers serving at transmission system voltages.

Optimal maintenance and asset management of high-voltage switching equipment require that information about their technical condition is available. For example, knowledge about the condition of important parts, with regard to contact wear, is useful when planning maintenance. The methods used to obtain such information are generally referred to as *diagnostic techniques*. This chapter describes in some detail the most commonly used techniques. Knowledge about what types of failures and technical problems switchgear are most likely to experience during service constitutes important background information for deciding on what diagnostics tests to apply. Therefore, a review of available failure statistics for switching equipment is given initially.

© Springer International Publishing AG 2017
K. Niayesh and M. Runde, *Power Switching Components*, Power Systems,
DOI 10.1007/978-3-319-51460-4_5

5.1 Reliability Surveys

5.1.1 Organization, Approach and Participation

The International Council on Large Electric Systems (CIGRÉ) [1], an international non-governmental organization, has over the years—among many other things—organized several international surveys on service experience with high voltage apparatus and systems, including switching equipment.

In the switchgear surveys, electric utilities from all over the world have participated by providing information about failures that have occurred on their equipment during normal service. Data are collected about the failed components (technology, age, voltage rating, switching duties, etc.) together with details describing the failures themselves, such as origin and cause, what sub-assembly failed, whether this was a minor or major failure, if environmental stress contributed, etc. The information is then analysed statistically, and the results together with recommendations are presented in extensive open reports. Of particular interest is identifying the most common failure types and origins, and how the failure frequencies vary with the various design and operational parameters, and with age. Obviously, when considering what types of diagnostic tests that are appropriate for a breaker population, and when planning the type and frequency of maintenance, such failure statistics are very useful.

Another objective of the surveys is to identify trends in reliability and service experience by comparing the findings from one survey with those from previous ones. To be able to do so in a trustworthy manner, central definitions and questions in each survey must be completely, or at least nearly identical. In particular, failure definitions are very important. For the most severe and thus most interesting failures, the term *major failure* (MaF) as defined in the IEC circuit breaker standard [2] has been adopted. A switchgear major failure is defined as "failure of a switchgear and control gear which causes the cessation of one or more of its fundamental functions. A major failure will result in an immediate change in the system operating conditions, e.g. the backup protective equipment will be required to remove the fault, or will result in mandatory removal from service within 30 min for unscheduled maintenance".

Table 5.1 shows some key parameters concerning the participation and the time of the three circuit-breaker surveys that CIGRÉ has carried out. All three surveys covered voltage classes from around 60 kV and above. Hence, medium voltage equipment, such as vacuum circuit breakers, is not included.

The third survey is by far the most comprehensive, as it comprises more than 280,000 years of circuit-breaker service. That is, information about all failures occurring in the four-year period 2004–07 in a circuit-breaker population of some 70,000 three-phase units is supposedly recorded. However, such surveys will always be associated with a certain under-reporting; a considerable portion of the failures that actually happen are not registered for various reasons.

Table 5.1 International circuit-breaker service experience surveys carried out by CIGRÉ

	First survey [3]	Second survey [4]	Third survey [5]
Time period	1974–77	1988–91	2004–07
Circuit-breaker technologies included	All types	Single pressure SF$_6$	Single pressure SF$_6$
Participation	120 utilities from 22 countries	132 utilities from 22 countries	83 utilities from 26 countries
Surveyed service experience	77,892 circuit-breaker—years	70,708 circuit-breaker—years	281,090 circuit-breaker—years

The 2004–07 survey also included disconnectors and earthing switches (all technologies). For switching equipment at distribution level voltages no international reliability surveys of an extent comparable to the three investigations mentioned above exist. Consequently, convincing failure statistics for such components based on information from a large number of end user responses are still lacking.

5.1.2 Results

5.1.2.1 Failure Frequencies

Major failure frequencies as a function of voltage level are plotted in Fig. 5.1. When considering the total populations, the failure frequency was about 1.6, 0.7 and 0.3 failures per 100 circuit-breaker—years in the first, second and third surveys, respectively. Thus, a circuit-breaker population of 100 units with the overall failure rate of the third survey can expect in average one major failure to occur approximately every three years.

Fig. 5.1 Major failure frequency as a function of voltage level for the 1974–77, 1988–91 and 2004–07 circuit-breaker surveys

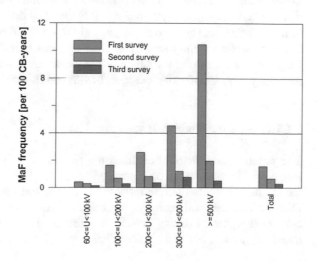

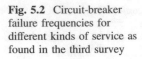
Fig. 5.2 Circuit-breaker failure frequencies for different kinds of service as found in the third survey

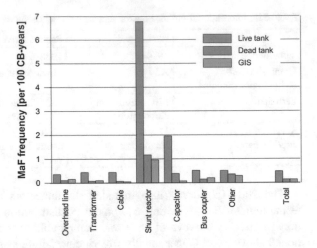

The most striking observation is that failure frequencies have decreased drastically in the 30 years passed between the first and third survey. For example, circuit breakers operating at system voltage range from 300 to 500 kV, which includes the widely used and important 420 kV level, had six times as many failures in the first survey as in the third. The much higher failure frequency at higher voltages in the first survey may be attributed to early, pioneering designs and design immaturity at that time. Later, the technology has matured, and the general reliability of the circuit breakers have improved substantially. However, even in the third survey the failure frequency increases with increasing voltage level.

The third survey correlates failure frequencies to the breaker's kind of service; i.e., whether it switches overhead lines, cables, transformers, reactors, etc. The results are shown in Fig. 5.2.

The failure frequencies for circuit breakers operating on shunt reactors (i.e., interrupting small inductive load currents) are around one order of magnitude higher than for line and transformer breakers. It is assumed that this can be attributed partly to the fact that the former ones in general are operated more frequently, and partly because the interrupting small inductive currents is a demanding switching duty. Capacitive load current switching is also associated with a clearly higher failure frequency than transformer, cable and overhead line switching.

5.1.2.2   Failure Modes, Origins and Causes

The failure mode distribution from the third survey is shown in the left part of Fig. 5.3. "Does not close on command" and "Locking in open or closed position" are the most common failure modes. Hence, the most frequently occurring serious failure on circuit breakers is that nothing happens when it is given an open or close command signal. "Locking" means that the circuit breaker is—for some reason— unable to switch, but that an alarm has been triggered by the control system.

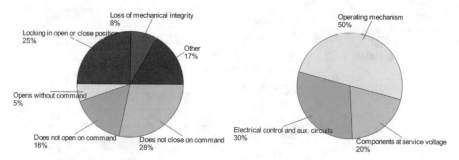

Fig. 5.3 Distribution of circuit-breaker failure modes (*left*) and sub-components responsible for the failure (*right*) as found in the third survey

When identifying the sub-component responsible for the failures, the distribution becomes as shown in the right part of Fig. 5.3. Roughly, half of the failures occur in the operating mechanism. Hence, the springs, the hydraulic or the pneumatic systems and the mechanical linkage and other moving parts that rapidly bring the main contacts between open and closed positions are obviously critical to the reliability of the breakers.

Around 30% of the failures are caused by malfunctions in the electrical control and auxiliary circuitry. A particularly critical function turns out to be the conversion of the electrical command signal for open or close to a mechanical movement that releases the main contact so that it starts moving. Typically, the command signal goes to a coil/solenoid which in turn releases a latch or opens a hydraulic or pneumatic valve.

Only one fifth of the failures occur on components at service voltage. Thus, what can be referred to as the "high voltage" parts of the switchgear, such as contacts and insulators are quite reliable. Dielectric breakdowns and failed interruptions are rare events. This suggests that the type tests that such circuit breakers have to pass are sufficiently demanding.

The third survey also asked for the utilities to identify the primary cause for failures. Table 5.2 shows the distribution of the responses.

Almost half of the failures reported were attributed to wear and aging. Apart from this, no other cause stands out. The fact that in nearly 25% of cases was "Unknown" or "Others" may suggest that identifying the primary cause for the failures is not always easy.

Concerning switching equipment other than circuit breakers, for example disconnectors and earthing switches, the most important findings are the same. The majority of the failures occurs in the release mechanism or in the moving parts, which move the main contacts between open and closed position. The cause of the failure is very often reported as "wear/aging".

The three circuit-breaker surveys also recorded minor failures, i.e. flaws and problems not immediately affecting the breakers ability to operate. Here leaks are by far the most common failure type. For circuit breakers, small SF_6 leaks from the

Table 5.2 Distribution of primary cause for failures as found in the third circuit-breaker survey

Primary cause	[%]
Design fault	5.8
Manufacturing fault	7.6
Incorrect transport or erection	3.6
Lightning overvoltage	1.8
Mechanical stress	1.3
Wear/ageing (incl. corrosion)	46.7
Incorrect operation	1.1
Incorrect monitoring	1.7
Mechanical failure of adjacent equipment	1.3
Incorrect maintenance	3.3
External damage	1.0
Others	8.3
Unknown causes	16.5

arcing chambers are very common. Moreover, minor oil leaks from the high-pressure hydraulic operating mechanism were also reported as a frequent problem.

5.2 Diagnostic Techniques

5.2.1 General

A wide variety of diagnostic techniques exist for switching equipment, from very simple readings of already available information—such as the accumulated number of switching operations—to complicated computerized systems for continuously monitoring of e.g. partial discharge activity.

The state-of-the-art of this field will be reviewed in the following. Emphasis is put on the technical principles and on the parameters monitored. The properties, advantages and disadvantages of the various methods and sensors are discussed. The description of sensors for measuring time, position, voltage, current and other common physical quantities is in most cases limited, as detailed surveys of the multitude of available sensors and methods can be found elsewhere.

An essential matter in the context of diagnostic testing and monitoring is the interpretation of the measurements. In some cases, it is straightforward to convert measurements into useful information about the condition of the equipment, in other cases it is not. Hence, the significance of the measurements needs to be thought carefully about before data is collected. Criteria determining whether equipment is "healthy" or not must exist. A major challenge is establishing criteria, which are generally valid. Quite often, a correct evaluation of measurements requires detailed knowledge and experience with the type of switching equipment under consideration. For example, a deviation in operating time of 2–3 ms from one year to the next may be within the normal variations for some circuit breakers, while for others this is a reliable early warning of problems with the operating mechanism.

Thus, diagnostics on switching equipment is far more than just collecting data. As in most other aspects of handling technical equipment, the users must know what they are doing and why it is being done.

The tasks of a switching device can be grouped into five primary or basic functions: (i) insulation, (ii) current carrying, (iii) switching, (iv) mechanical operation, and (v) control and auxiliary functions. In the present overview, the diagnostic techniques and parameters are categorized according to which of these main functions they cover.

Only techniques and sensors that are commercially available and widely used are included. Furthermore, the review also reflects the dominant market share of single pressure SF_6 circuit breakers and the large utility interest in diagnostic techniques for this breaker technology.

Most of the information provided below can also be found in a CIGRÉ Technical Brochure [6]. More detailed treatment of other aspects of diagnostics on switching equipment, such as justification, test requirements, system architecture, and information management can be found in the same source.

5.2.2 Insulation

5.2.2.1 Failure Mechanisms

The electric insulation function in switching equipment is provided by a combination of gaseous, liquid and solid dielectric materials. A complete insulation failure during service usually leads to a spark-over, either between phases, between phase(s) and ground, or across an open switching device. The breakdown may occur directly through the insulating gas or liquid, along the surface or through the bulk of solid insulators.

A spark-over in a switching device is always a dramatic incident, whether it occurs internally or across the device terminals. In most cases, the equipment is severely damaged, and major repairs or replacements are required.

A certain percentage of the reported major failures with switching equipment are attributed to the dielectric system. However, several other faults and irregularities, such as incorrect contact positioning, breakdown of mechanical parts, excessive voltage transients and contact overheating may as a secondary effect lead to an insulation failure. Due to the heavy damage caused by a spark-over, the primary cause for the failure may in several cases not be found, and it is reasonable to assume that the dielectric system is, incorrectly, blamed also for a portion of failures of other origins.

The most frequently used techniques and sensors used for periodic (P) diagnostic testing and/or continuous (C) monitoring of parameters related to the basic function "insulation" are summarized in Table 5.3, and will in the following be reviewed in more detail.

5.2.2.2 Quantity of Insulation Medium

An obvious condition that must be satisfied to maintain the specified dielectric integrity of a switching device is that the insulation is present in sufficient amounts. For SF_6 and oil insulated equipment, respectively, this means that SF_6 density and oil level are being held above a safe operating minimum value.

SF_6 gas density is usually measured by using a temperature compensated pressure gauge. Although systems for remote reading of gauges exist, gauges are most frequently used in combination with a switch to give a signal if the density drops below predetermined limits. Gas density is a very important parameter, and virtually all SF_6 breakers are equipped with systems that give alarms and/or block for operation if the gas density is too low.

SF_6 leaks can also be detected directly. The available methods span in complexity from using soapy water, through handheld gas detectors ("sniffers") to SF_6-sensitive imaging systems.

A sniffer detects presence of SF_6 in air by measuring a reduction in the electric conductivity of the air. The imaging system contains a laser illuminating the SF_6 insulated equipment with infrared light at a wavelength that SF_6 strongly absorbs and a camera that is sensitive for the same wavelengths. An SF_6 leak is observable as "black smoke" against a lighter background. The great advantages with this system are high sensitivity (a detection limit as low as leaks of 1 kg SF_6 per year is claimed), and that it can be used without taking the switching equipment out of service. The main drawbacks are that the imaging system is bulky and expensive, and that wind can make it difficult to do outdoor measurements.

In minimum oil and bulk oil circuit breakers, the quantity of insulation oil is usually indicated with oil level indicators of various types. Mechanical indicators (float level devices) and observation windows are usually installed. These are for periodic visual readings, and require that a person inspects the individual units. Observation windows have the disadvantage of easily becoming stained by oil after a few breaker operations, and may thus give the impression of a correct oil level, even if there is no oil in the tank.

In vacuum interrupters, metallic arcing debris or loss of vacuum may impair the electric insulation. The most common method to check the dielectric integrity of vacuum "bottles" is by AC or DC high potential testing. A voltage a few times the rated voltage is applied, and a high leakage current indicates that the insulation system has deteriorated. X-ray generation during testing is extremely low with recommended voltage and normal contact spacing.

Table 5.3 Diagnostic techniques and sensors for testing the insulation of switching equipment

Parameter	Application(s)	Method/sensor	
Quantity of insulating medium	SF_6 D, ES in GIS	SF_6 density by:	
		Temperature compensated pressure gauge	C
	SF_6 D, ES in GIS	SF_6 leak detection by:	
		Soapy water	P
		Sniffers	P
		Infrared imaging system ("thermovision")	P
	Oil	Oil level by:	
		Mechanical, optical or electronic level indicator	C, P
		Observation window	P
	Vac	High potential insulation testing	P
Quality of insulation	All	Partial discharges by:	
		UHF	C, P
		Acoustics	P
	All	Insulation resistance	P
	SF_6 D, ES in GIS	Moisture content by:	
		Dew point	P
		Solid state sensor	P
	SF_6 D, ES in GIS	Purity of SF_6 by:	
		Chemical detection tube	P
		SF_6 content	P
	SF_6 D, ES in GIS	Gas temperature measurement to verify that the SF_6 remains above its condensation point	C
	Oil	Oil quality by:	
		Moisture content by coulometric titration	P
		Dissolved gas analysis	P
		Acidity measurement	P
		Oxygen content	P
Insulation distance	All	Contact position transducer	C, P
	D, ES	Visual inspection	P

Abbreviations
SF_6 SF_6 circuit-breakers
Oil minimum oil and oil tank circuit-breakers
Vac vacuum interrupters
ES earthing switches
D disconnectors
GIS gas insulated substations
C used for continuous monitoring
P used for periodic diagnostic testing

Voltage sources for testing of vacuum breakers are commercially available, and this is a fast and reliable periodic diagnostic test, provided that the vacuum bottles are easily accessible.

5.2.2.3 Quality of Insulation

The dielectric quality of the electric insulation is affected by several factors and is, therefore, in general harder to determine than the quantity of the insulation.

Some of the diagnostic methods or sensors for measuring the quality of the insulation are general in that they consider parameters that reflect the overall condition of the insulation system of the device. An example of such a technique is partial discharge measurements. Other methods or sensors measure specific physical quantities of the insulating medium, such as a moisture content or dissolved gases in oil. Consequently, some techniques apply to all types of switching equipment, while others are useful for one or a few dielectric materials only.

Major insulation failures (spark-overs) in gaseous, liquid or solid insulation are usually preceded by partial discharges. Thus, information about discharge activity is very useful when assessing the quality of the insulation. Electrical methods for measuring partial discharges are however, in general not applicable on equipment in normal service, as it is very difficult to obtain a sufficient sensitivity in the noisy environment of a substation. However, one exception exists. In gas insulated substations (GIS) where the insulation is surrounded by a grounded metallic enclosure, sensitive systems for continuous monitoring of the discharge activity by recordings in the ultra-high frequency (UHF) range are commercially available. A significant number of antennas have to be installed inside the GIS, and a comprehensive data handling system is required for continuous monitoring. However, in some GIS designs in which discontinuities in the metallic enclosure exist (e.g. observation windows with a minimum diameter of around 5 mm, or support insulators with no external metallic ring), UHF antennas can be installed non-invasively. Nevertheless, UHF monitoring is rather expensive, so retrofitting or installation in other than very important stations can in most cases not be justified.

Partial discharges inside a GIS also generate acoustic signals (sound waves). The signals propagate through the SF_6 and into the metallic enclosure and can be detected by using sensitive acoustic sensors mounted externally on the enclosure. Furthermore, experience has shown that small particle contamination inside the GIS ducts can reduce the quality of the insulation system significantly by initiating spark-overs. The strong electric field causes the particles to move around or "jump", and their impacts in the enclosure can also be detected from the outside with acoustic sensors.

Consequently, acoustic methods have been developed for detecting and characterizing discharges and particles in GIS. The sensitivity depends greatly on the distance between the source and sensor, and is in general good as long as there is no flanges between. For example, moving particles of size less that one millimetre are easily detected at a distance of 1 m.

Commercial instruments for periodic diagnostic testing using the acoustic method is available. With regard to continuous monitoring, the same considerations apply as for the UHF method; the large number of sensors and the comprehensive data management system make it difficult to justify this economically. Hence, diagnostic testing using a portable instrument may be an economic alternative.

Another technique that considers the overall dielectric system is insulation resistance measurement. This parameter can easily be measured with a portable instrument (a "megger"), provided that the equipment is disconnected. Excessive surface contamination on solid insulators may reduce the insulation resistance, and can be detected this way. Periodic measurements of the insulation resistance may thus give some indications if the insulation is degrading. However, the opposite is not necessarily true; a high insulation resistance does not guarantee that the insulation system is in a good shape. Consequently, insulation resistance measurements are best suited as a supplement to other diagnostic methods.

The dielectric properties of SF_6 are in general not very sensitive to the quality and purity of the gas. It is however very important to keep the content of moisture low, not because moisture in its gaseous state has a dramatic effect on the dielectric performance of the gas, but because temperature fluctuations may cause the moisture to condense. The resulting water droplets on solid insulators may, in combination with other impurities, reduce the dielectric withstand significantly. Consequently, moisture-adsorbing scavengers are commonly used, and most manufacturers have guidelines for the maximum allowable levels for moisture in their equipment.

For moisture measurements in gaseous insulation, two types of instrumentation are used. The first type is based on determining the dew point of the gas, i.e. the temperature at which the vapour condenses as liquid or frost. The second type employs sensors with hydrophilic materials. When the moisture content changes, this can be measured as a change in the electrical resistance, capacitance or extent of electrolysis in the sensor.

Portable equipment with sampling capability is commercially available for periodic measurements.

Also for a liquid insulation as oil, condensation of moisture to water droplets may reduce the quality of the insulation. Thus, the moisture content is an important parameter in some oil-insulated switching equipment. For other equipment designs, including the so-called free-breathing oil circuit-breakers, in which oil is always saturated with humidity, the moisture content in oil is of course not critical, but the solid insulation inside the circuit-breaker may be contaminated if there is a major water infiltration.

Moisture content in an oil sample is most commonly determined using coulometric Karl Fischer titration. This is an accurate and reliable method, but only suited for periodic testing.

A more general test on a sample of oil insulation is to measure the dielectric strength directly. The standard specifies the geometry of the test cell and the test procedure. Provided that the sample is representative for the oil in the device, this test is probably the best way to check the quality of the insulation oil. However, after a small number of operations, the dielectric strength of oil is strongly affected, but this is in most cases not critical for the circuit breaker.

Several other chemical methods that are applied for assessing the condition of transformer oil (dissolved gas analysis by chromatography, acidity measurements, measurements of oxygen content, etc.) can in principle also be used for oil-filled switching equipment. However, a few important matters should be kept in mind when considering these chemical methods: First, the oil in switching devices will inevitably contain dissolved gases generated by the arc, while arcing products in transformer oil is a clear indication of severe problems. Second, the insulation system in a transformer includes paper, and this causes the degradation mechanism in several respects to be different from a paper-free oil system. Third, equipment for continuous monitoring of transformers will in most cases be too expensive to be considered for oil-insulated switching equipment. Thus, when applied on breakers, these chemical methods are practical for sampling and periodic testing only.

5.2.2.4   Insulation Distance

Malfunctions or incorrect adjustments in the operating mechanism or in the mechanical linkage between the operating mechanism and the moving contacts can render the primary contacts in incorrect end positions after an operation. If the contacts do not become fully separated, the insulation distance can be reduced, thereby lowering the dielectric withstand level across the switching device.

This is a serious condition, and most breakers are equipped with auxiliary contacts that give alarms immediately if the mechanical travel is not completed during a switching operation. In most circuit breakers, these auxiliary contacts are mechanically linked to the operating mechanism, not directly to the moving contact itself. This can be a disadvantage, since a correct notification then presupposes that the mechanical linkage between the operating mechanism and the contact is functioning as intended and correctly adjusted. For example, if a shaft ruptures during an opening operation, the contact may still be closed even if the operating mechanism and thus the auxiliary contacts indicate that the breaker is open.

Consequently, auxiliary contacts or other types of position transducers (optical, mechanical, electronic) for indicating whether the insulation distance in an opened contact is correct, should be installed as close to the moving contact member as possible. However, the demanding environment in and nearby the primary contacts (electric disturbances, arcing products, high temperature and pressure) requires that the sensors are rugged and robust.

In many cases, especially in earthing switches and disconnectors, the position of the contacts is directly observable. In these cases, the insulation distance can be checked on a periodical bases by a simple visual inspection.

5.2.3   Current Carrying

5.2.3.1   Failure Mechanisms

The current carrying ability of the primary circuit of a switching device in the closed position is crucial to its reliability. Failing to fulfil this function usually manifests itself as overheating in contacts, connectors, joints or other parts of the primary circuit. This may over time lead to contact welding, severe deterioration of the insulation system or other major failures.

Experience has shown that as long as the contacts in switching equipment are not subjected to other operating conditions than they are designed for, a typical contact degradation process usually develops rather slowly in the beginning. It may take several years to reach the rapidly self-accelerating final stages that eventually can lead to a complete breakdown. Thus, most contact problems can be disclosed by periodic diagnostic testing, provided that the applied diagnostic technique is sufficiently sensitive. Only minor advantages can be expected by going from periodic testing to continuous monitoring of the current carrying ability of the switching device, at least with regard to detecting contact degradation.

Table 5.4 summarizes features of a number of diagnostic techniques and sensors for monitoring the basic function "current carrying".

5.2.3.2   Contact Resistance

Measuring the electric resistance in the primary circuit is probably the mostly applied method for evaluating the current carrying capability. An increasing or fluctuating resistance, or large deviations between the values obtained from similar

Table 5.4 Diagnostic techniques and sensors for testing the current carrying ability of switching equipment

Parameter	Application(s)	Method/sensor	
Contact resistance	All	Four point resistance measurement	P
Temperature of contacts and breaker unit	All	Infrared imaging	P
		Temperature at a point by:	
		Thermocouple	C, P
		Optical sensor	C, P
		Infrared sensor	C, P
Contact penetration	All	Contact position transducer	C

Abbreviations
SF_6 SF$_6$ circuit breakers
C used for continuous monitoring
P used for periodic diagnostic testing

components (e.g., between the three phases) are clear indications of abnormalities in the current carrying properties of a device.

Resistance measurement is carried out by disconnecting the switching device from the high-voltage system and then passing a measuring current from a separate source, typically a portable rectifier or a battery. The resistance is determined by the four-point method, i.e., by using separate connection points for the current supply leads and for the voltage probes, see Fig. 5.4.

The current source should provide DC as passing an AC current introduces a time-varying magnetic flux surrounding the current path. This induces an additional voltage difference between the voltage probes, yielding (often grossly) misleading resistance values.

The resistance values obtained for modern switching devices, e.g. SF_6 circuit breakers with silver plated main contacts, are typically in the range 10–100 $\mu\Omega$ (Normal handheld multifunction meters have an insufficient sensitivity for this application). To obtain a voltage drop sufficiently high to determine the resistance with reasonable accuracy, it is recommended to pass a current of substantial magnitude. Moreover, thermoelectric effects may not be negligible in cases where both resistance and measuring current are low. Thermoelectric effects can however be compensated for by taking the average resistance value from two measurements carried out with different polarity.

Furthermore, contact grease, various decomposition products and other compounds present at the contact surfaces have occasionally been found to lead to a resistance that is not independent of the magnitude of the current. Lower values are obtained when a high current has "warmed up" or "baked" the surface films.

Thus, doing resistance measurements using currents orders of magnitude below the rated load current of the equipment may give misleading results. For typical switching devices with a rated load current of a few kilo amperes, it is for these reasons, normally recommended to use measuring currents of at least a few hundred amperes.

Fig. 5.4 Four-point method for measuring resistance

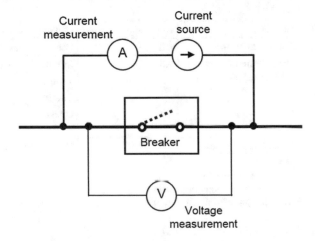

Resistance measurements require that the high-voltage conductors on both sides of the switching device can be directly accessed. If one side can be grounded near the switching device, it is sufficient to have access to the other side and to ground. This has a couple of important implications when considering the applicability of this diagnostic technique. First, the power component has to be taken out of service while doing measurements, so this method is only usable for periodic testing.

Second, the method is less suited for equipment in gas-insulated substations (GIS) and in other cases where more or less dismantling work is required before the current source and voltage probes can be connected across the switching device. A measurement over several components connected in series, for example between the outdoor termination of an overhead line and the busbar earthing switch of a GIS, becomes less sensitive. A significant resistance increase in one of the circuit breakers, disconnectors or busbars along this section may constitute only a small fraction of the total resistance measured, and can thus easily be overlooked.

Still however, resistance measurements are in many cases a simple, direct, fast and reliable method for verifying the current carrying ability of a switching device.

5.2.3.3 Temperature

Overheating due to faulty conditions in the primary circuit can also be detected by thermal methods, such as infrared sensors and imaging systems, thermocouples and optical fibre measurements.

Cameras imaging infrared radiation ("thermovision") are used for detecting poor contacts and other parts of substation equipment that are abnormally heated. Such advanced thermographic equipment is employed exclusively for periodic diagnostic testing, and the power components to be tested have to be in service and preferably under high load. Figure 5.5 shows an example of how abnormally heated contacts appear in an infrared imaging system.

The sensitivity of infrared imaging systems is usually good, both in terms of geometrical and thermal resolution. Normally, deviations in surface temperatures as low as a one degree centigrade can be revealed. Thus, if a deteriorating contact is situated in such a way that it is directly observable through the infrared camera, it is possible to detect the abnormality at an early stage. Disconnectors and overhead line terminations at the circuit breakers of an air-insulated outdoor substation are among the switching components where incipient problems in the current carrying condition can be detected with an infrared camera. Moderate overheating of less accessible parts, like the main contacts in circuit breakers inside a GIS, are significantly more difficult to reveal. Here the heat spreads out in the conductor, in the insulation system and in the enclosure, resulting in a far lower temperature rise on the outer surface of the enclosure than in the contact. In addition, heat generated by ground currents in the enclosure may further disguise an overheating in the high-voltage primary circuit.

Disturbances from other infrared radiation sources can complicate in some cases the use of infrared cameras. When surveying outdoor substations in clear weather,

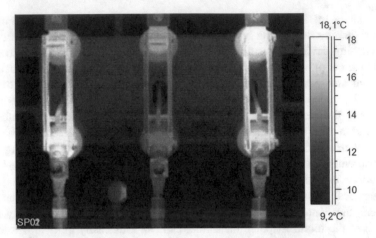

Fig. 5.5 Infrared image of a three phase 16 kV disconnector. The overheated contacts and terminations of the *left* and *right* units are clearly seen as light areas. The temperature span within the frame of the image is shown on the *bar* to the *right*

sunlight reflecting on the equipment may give readings corresponding to the surface temperature of the sun rather than the surface temperature of the equipment. For this reason, infrared camera surveillance is most conveniently carried out under overcast weather. When operating indoor, light bulbs and tubes may cause similar interference. For an experienced operator however, such disturbances are rather easily handled and not considered as a major limitation of the method.

The infrared emissivity of different surfaces and materials varies quite a lot. Consequently, determining the temperature on a surface on an absolute scale with an infrared camera is not possible without knowing the emissivity. Emissivity tables and calibration methods exist, but in general, it is considerably easier to demonstrate overheating as temperature deviations than to measure the absolute temperature accurately.

Absolute temperature measurements are however important in evaluating how critical a defect is, for comparing with earlier measurements at the same component, as well as for relating a measurement to the temperature specifications given by the manufacturer of the component. Such evaluations should preferably be based on measurements obtained under full load. If they are not, the effect of load on heat generation and temperature rise must be taken into account, and this may introduce significant uncertainties. Furthermore, when doing outdoor measurements, the ambient conditions such as wind speed and temperature will lead to further inaccuracies, and interpretation and evaluations become correspondingly less certain.

Infrared imaging systems have been extensively used by many utilities for several years. Infrared cameras have gradually become more compact and easier to use, but they are still rather expensive.

In general, this is found to be a fast and reliable diagnostic technique, provided that the operator has some experience and is aware of its limitations. Moreover, infrared measurements are completely non-invasive and do not interfere with the normal service of the substation.

Several sensors for temperature measurements at a point exist, including thermocouples, infrared point sensors and optical fibre point sensors. Optical fibre sensors are all dielectric and can in principle be attached directly to live parts, but so far, this has not been applied to any notable extent. An obvious disadvantage with point sensors is that in many cases a great number is required to cover a component properly, resulting in a lot of wiring and a comprehensive data handling system. Most types of point sensors can only be applied on grounded parts. This limits their application to measuring the overall temperature of the switching device, rather than the temperature at the primary high-voltage conductor. The sensitivity for detecting overheating is less, and measurements are more influenced by external factors (e.g. ambient temperature) than with direct measurements on the primary conductor.

5.2.3.4 Contact Position

Usually, switching equipment has position transducers installed to indicate whether the primary contacts are in closed or open position. These are primarily used as status indicators for system operation purposes. However, such position sensors may also provide more detailed information on the current carrying ability of a switching device by indicating whether the contacts penetrate into each other, as they should. In circuit breakers with separate arcing and main contacts (e.g. single pressure SF_6 circuit breakers), insufficient penetration may cause the current to pass through the arcing contacts only. The arcing contacts are not designed for continuous operation and will overheat within hours, leading to severe problems.

The various methods and sensors used to monitor the position and travel of contacts will be extensively treated in Sect. 5.2.4.3.

5.2.4 Switching

5.2.4.1 Failure Mechanisms

Depending on the type of switching equipment, its location in the network, the network topology and the type of switching event, the dielectric, thermal and mechanical stress during a switching operation can vary over a wide range. Furthermore, making or breaking high currents means handling of large amounts of electric power, so unsuccessful switching may cause the breaker to fail catastrophically within a few power cycles. Consequently, to ensure a high reliability, each type of switching equipment has to pass through several switching tests, and fortunately, switching failures are rare.

Most of the great number of possible failure types do not appear without any previous notice. The diagnostic techniques, summarized in Table 5.5, together with type test and field experience can help in detecting abnormal conditions and prevent failures in the basic function "switching".

5.2.4.2   Position of Primary Contacts

Both from network operation and from the control system point of view, the positions of the contacts of switching equipment are very important information. Consequently, nearly all types of switching devices are equipped with position transducers that continuously monitor whether the contacts are in open or closed position. A number of failure types, e.g. problems with the operating mechanism, can be detected as a missing notification of that the moving contact has reached its final position.

For switching equipment with the moving contacts mechanically connected to an operating mechanism at ground potential, the end positions ("open" or "closed") can be monitored rather simply by electronic, optical or electromechanical sensors in the operating mechanism. The most commonly used position sensor type is an electromechanical auxiliary switch connected directly to the operating mechanism or to some grounded part of the mechanical transfer system.

Various types of electronic proximity sensor are also applied. Proximity sensors operate without contact and respond on geometrical changes in the immediate vicinity of the sensor. Consider for example, a metallic latch being positioned in front of a proximity sensor containing an electric circuit with a coil and a capacitor. The characteristics of the oscillations in this LC-circuit (frequency, damping, etc.) will change because the latch introduces a change in the magnetic circuit seen by the coil.

The position of a moving part can also be sensed by optical systems, for example by acting as a barrier between an optical source and an optical sensor. Reflectors can be employed to reduce the number of active components in the system. The main advantages with optical systems are that they are immune to electrical disturbances, and due to the insulating properties of optical fibres, they can even be considered for use on live parts. However, optical position transducers have so far not been extensively applied in switching equipment.

As discussed in Sect. 5.2.3.4, it should be kept in mind that in most cases contact position is measured indirectly. Consequently, failures or incorrect adjustments in the mechanical linkage between the contacts on high-voltage potential and the position transducer on ground can give misleading readings.

5.2.4.3   Contact Travel Characteristics

The terms travel characteristic/travel curve denote measurements of the position of the primary contacts as a function of time during operation of a switching device.

Table 5.5 Diagnostic techniques and sensors for testing the switching function of circuit breakers and other types of switching equipment

Parameter	Application(s)	Method/sensor	
Position of primary contacts	All	Position transducer, e.g.:	
		Auxiliary switch, contactor	C
		Electronic proximity sensor	C
		Optical sensor	C
Contact travel characteristics (position, velocity, acceleration)	All	Dynamic position sensor with analogue output:	
		Resistance potentiometer	C, P
		Magneto-resistive sensor	C, P
		Linear variable differential transformer	C, P
		Dynamic position sensor with digital output:	
		Optical with incremental coding	C, P
		Optical with non-incremental (absolute) coding	C, P
Operating time	All	Electrical recording of time to close/open of primary circuit	P
	ES, D	Motor running time	C, P
Pole discrepancy in operating times	All	Electrical recording of time to close/open of primary	C, P
		Circuit	
	ES, D	Motor running time	C, P
Arcing contact wear	CB	Accumulated $I^2 t_{arc}$ by:	
		Current and time measurement	C
		Statistical estimates (FAO)	P
	All	Dynamic contact resistance	P

Abbreviations
CB SF_6 minimum oil and oil tank circuit breakers
ES earthing switches
D disconnectors
C used for continuous monitoring
P used for periodic diagnostic testing

Contact travel characteristics can easily be recorded on most types of switching equipment, also during service. Some exceptions do however exist. In some types of switching equipment, the moving contacts are difficult to access, and this complicates the recordings. Moreover, the concerns related to that the position transducers in most cases are not attached directly to the contacts, but on some location between the operating mechanism and the contact, are relevant also in this context.

Knowing the contact position as a function of time, contact velocity and acceleration can be derived. These parameters are also used for monitoring purposes. Figure 5.6 shows examples of determined contact position, velocity and acceleration as a function of time related to an opening operation of a typical circuit breaker.

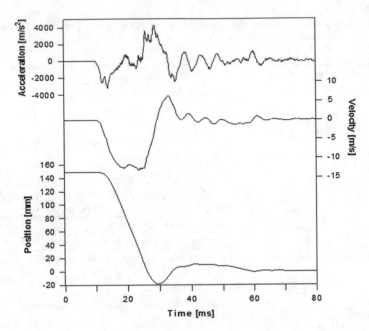

Fig. 5.6 Example of normal travel characteristics from an opening operation of a circuit-breaker. Velocity and acceleration are found as first and second derivatives of the position versus time measurement. The contacts separate after around 22 ms

Travel curves contain information on several important aspects of the mechanical operation of the contacts, and consequently, a number of abnormalities and failure mechanisms that may occur can be detected from a careful analysis of travel curves. Figure 5.7 shows some schematic examples.

Travel curve sensors are essentially position transducers, but in contrast to the contact position sensors discussed in the previous section, travel sensors measure the position of the contract throughout a closing or opening operation. Travel sensors are usually categorized according to whether the output signal is on analogue or digital form.

Among the analogue sensors, potentiometers are the most commonly used. Potentiometers are available in both rotational and linear versions and give an analogue DC voltage output that is proportional to the linear or rotational movement. Potentiometers are inexpensive and reliable and have a sufficient accuracy when used correctly. When recording linear movement in a switching device the sensor typically need a measuring range of at least 5–20 cm. An exception is vacuum breakers, where the travel distance is of the order of 1–2 cm only.

The potentiometers are attached directly to a moving part of the switching device, usually in or near the operating mechanism. This makes them vulnerable to the severe mechanical shocks generated during a breaker operation; they may simply loosen.

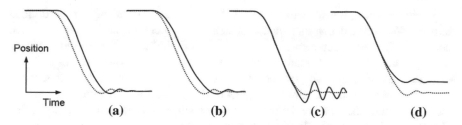

Fig. 5.7 Normal travel curves (*broken lines*) and schematic examples of abnormal contact travel recordings (*solid lines*) from an opening operation: delay in release mechanism, e.g. due to poorly lubricated release latches (**a**), low contact speed, e.g. due to reduced energy in operating mechanism (**b**), poor damping, e.g. due to defective dash pot (**c**), and too low insulation distance in open position, e.g. due to incorrectly assembly (**d**)

Magneto-resistive travel sensors work without contact and make use of angle measurement in an axisymmetric magnetic arrangement. A rotation is measurable when a magneto-resistive sensor is implemented in a field of rotatable pivoted magnet core or magnet system. The electrical resistance of a ferromagnetic layer depends on the angle between the rotatable magnet core and the internal magnetizing of the ferromagnetic layer. In a typical design, a magneto-resistive sensor consists of four sensor elements connected in a Wheatstone bridge. The output signal of this bridge is amplified and balanced and zero point and temperature compensation is carried out. The output signal can be current or voltage.

The linear variable differential transformer (LVDT) is a linear analogue non-contact sensor. In this electromechanical device, the output voltage is proportional to the axial displacement of a movable core. An LVDT consists of a primary winding and two secondary windings symmetrically located around an iron core. The secondary windings are connected in series, but wound in opposite direction. When energizing the primary winding with an external AC source, the output voltage is the difference between the induced voltages in the two secondary coils. These sensors are available for travel distances up to a few hundred millimetres.

Compared with potentiometer, magneto-resistive sensors and LVDTs are insensitive to mechanical tolerances and temperature variations. As a non-contact sensor type, they do not wear and are largely unaffected by dirt, contaminants or mechanical shocks. Sensor life is virtually infinite; they are capable of withstanding millions of cycles without failure. Thus, these sensors are well suited for application in a demanding environment as in a switching device.

A number of digital sensor types, both in linear and rotational designs are also available. Many designs are based on using optics. Typically, the movement is recorded when a ruler or disk attached to a moving part of the circuit breaker passes by an optical sensing system. The ruler or disc is equipped with one or more tracks of light barriers, reflectors, strip codes other types of digital codes that are recorded by the optical sensing system. The most sophisticated and expensive ones have a resolution approaching 10 μm. Optical systems have to be protected against contamination and dust.

Digital position transducers are all of the non-contact type and can be divided into incremental and non-incremental sensors. Incremental sensors give a relative position, while non-incremental systems apply a digital code (e.g. ordinary binary code or Gray code) that unambiguously determines the absolute position at any instant. Thus, also the end position of the contacts can be obtained accurately.

Incremental sensors require two tracks to determine the direction of the movement (whether the contact is opening or closing). Furthermore, to determine absolute position an additional track with at least one reference point must be applied. Non-incremental coding is more robust and reliable than incremental coding since in the case of a disturbance only single values become erroneous.

Travel sensors must be able to withstand the substantial mechanical forces and vibrations of a circuit-breaker operation. Furthermore, they should be protected from pollution and from electro-magnetic disturbances. Rotational travel sensors are easier to encapsulate than linear ones. Sensors that are permanently installed on the switching equipment must show a very high reliability over the entire lifetime of the equipment.

However, recording of travel curves for diagnostic purposes is still primarily applied on a periodic basis using temporarily mounted sensors, and the reliability requirements for the transducers are in several aspects less strict. For example, erroneous recordings due a loosened potentiometer are easily recognized, and a new recording can be made after securing the potentiometer.

Preferably, the position transducers should be connected directly to the moving contacts. In most cases, this is not feasible, as the contacts are not easily accessed. For example, on modern circuit breakers the moving contacts are inside the gas compartment. Alternatively, the sensors should be located as close to the contacts as possible. However, as discussed earlier, this introduces some uncertainty, as the measurements will be misleading if the mechanical linkage is incorrectly assembled or adjusted.

The normal procedure when doing periodic testing is to have one set of sensors and one data acquisition system that are moved from breaker to breaker. Hence, the time spent on attaching sensors becomes important. Occasionally, no easily accessible positions can be found, and travel curve measurement is not a viable diagnostic method, at least not when using temporarily mounted sensors. If sensors are permanently installed in the breaker, more time and effort in placing the sensors near the moving contacts can be justified.

When the travel sensor is mounted on some location between the moving contact and the operating mechanism, the recorded characteristic is not necessarily the travel curve for the contact. This is because the mechanical transfer (rods, gears, wheels etc.) in many cases is a non-linear system. Consequently, to determine the true contact movement (or speed or acceleration) conversion tables/transfer functions between the point of measurement and the contact must be applied. Figure 5.8 shows collected data from a rotational sensor and the linear travel curve determined using a conversion table.

Conversion tables are in general different from one breaker model to the other, and also unique for each sensor position. Usually, manufacturers of commercial

Fig. 5.8 Example showing conversion of data obtained with rotational transducer to linear contact movement

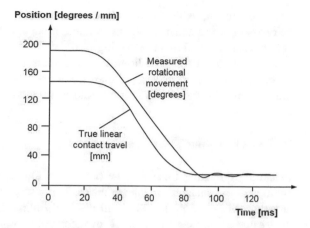

equipment for travel measurements specify where to attach the transducer and some provide conversion tables for the most common breaker models. However, for less common switching equipment, conversion tables may not be available. In these cases, the diagnostic testing has to be based on trend analyses and comparisons of the recorded characteristics. Specific values of parameters describing the contact travel, e.g. penetration distance and speed at the time of contact separation, cannot be obtained.

Although some of the characteristic features of a travel curve are similar for most switching devices of the same category, substantial variations can be found between different makes and models. Thus, a correct and detailed interpretation of travel curves requires some skills as well as knowledge and experience with the switching device being considered. However, as mentioned above, valuable information can be found simply by comparing travel curves from several breakers of the same type, although distinguishing normal, statistical variations from an abnormal one may not be straightforward. Some, but far from all breaker manufacturers provide information about how the contact is supposed to move (velocities at contact separation, timing information, etc.) and this can be very helpful in the interpretation of the measurements.

Furthermore, the travel characteristics of a circuit breaker may also in several ways be affected by conditions that not necessarily are considered abnormal. For example, a fully charged operating mechanism gives a faster contact movement than a partially charged mechanism. Similarly, recordings performed in cold ambient can differ from recordings obtained in a warmer ambient. When breaking high short-circuit currents with self-blast circuit breakers, the opening speed may be considerably lower than when breaking load currents. The reason is that in the high current case the arc for a short while completely blocks the gas flow through the nozzle.

Travel curve measurement is a broadly accepted and widely used diagnostic technique applied for periodic diagnostic testing on almost all types of switching equipment. Its greatest advantage is that it focuses on the contact movement, which is essential for the switching operation. In addition, it is sensitive to malfunctions in the operating mechanism, in the mechanical linkage between operating mechanism and contacts, as well as in the arcing chamber.

Test equipment for periodic testing has been available from several manufacturers for many years. Recently, some circuit breaker manufacturers have started to offer systems for travel curve measurements on a continuous basis.

5.2.4.4   Operating Time

Measuring operating times i.e. the time from the breaker receives the command signal until the contacts separate or make, is in a sense a simple alternative to travel curve recordings. Both methods will disclose timing deviations (delays), which in most cases are caused by problems of mechanical origin.

Accurate measurements of operating times are usually performed as a periodic check, and the switching device is disconnected from the high-voltage system. Typically this parameter is found by passing through a measuring current (a few amperes DC at a few volts) and determining at what time the voltage drop across the breaker changes from zero to the applied value or vice versa. The arcing contacts of some circuit-breakers types can have a rather high surface resistance, requiring a somewhat more powerful test signal. In circuit breakers with several arcing chamber in series, a separate current source and voltage recording is required for each chamber. An accuracy of one tenth of a millisecond is more than sufficient and easily obtainable.

For switching equipment in GIS, accurate measurements of operating times can be considerably more difficult than in air-insulated substations (AIS). Especially in older GIS cable bays, the primary circuit may be completely inaccessible, and operating times cannot be obtained non-invasively. However, most GIS manufacturers have for some years offered so-called insulated earthing switches as an option to facilitate timing measurements. With an insulated earthing switch, the current and/or voltage leads can be attached to the primary circuit through an (ungrounded) earthing switch.

An obvious disadvantage by using a measuring current from a separate source for determining operating times is that the switching device has to be taken out of service. An alternative approach that facilitates measurements on devices in service is to use a current path or branch in parallel to the branch of which the circuit breaker is a part of, see Fig. 5.9. Substations with 2 and 1.5 breaker systems, as well as most substations with bypass and ring arrangements can be configured this way.

The procedure is as follows: Initially, the breaker is closed so the load current is split between the two parallel branches. Then, the breaker is opened, and current commutes to the other branch as the contacts separate. Commutation takes place immediately, not at current zero as in an ordinary breaking operation. Thus, the opening time can be found by considering the output of the current transformer of the branch containing the breaker. Closing times can be found in a similar way.

Fig. 5.9 Circuit arrangement for measuring operating times without taking the breaker out of service (see text)

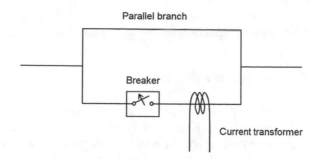

For circuit breakers with several contact sets in series, this method is less efficient as it does not allow for measuring the timing of the individual sets.

Without the parallel branch, timing measurements, which are based on the current transformer output, will determine when current starts flowing or breaks, not when the contacts separate or mate. Thus, during opening operation the arcing time will be included and yield a somewhat longer time, while pre-striking during closing causes the recorded time become somewhat shorter. Furthermore, variations with regard to how the current and voltage waveforms match the contact movement introduce statistical scatter in the measurements of around one half cycle (4–5 ms).

Time measurements other than opening and closing times can be used for diagnostic purposes of the switching function. These include arcing time, break time, time to arcing contact separation/touch, pre-arcing time, but do however, require more complex sensors and data acquisition and processing systems.

For switching equipment where contact movement is directly driven by a motor (e.g. disconnectors, earthing switches) then the motor runtime is a measure for the overall operating time.

Most modern circuit breakers show very small variations in their operating times. For example, closing and opening times of a spring-operated circuit breaker may consistently vary with less than 1–2 ms over years. For circuit breakers with hydraulic or pneumatic operating mechanism larger variations are normal, especially when comparing measurements obtained on outdoor circuit breakers in cold and warm weather. Thus, the criteria for distinguishing between normal and abnormal timing deviations must be based on experience and knowledge of the device being considered. Some, but not all manufacturers include operating times with tolerances in the specifications that come with the circuit breaker. This is certainly a great advantage for an inexperienced user when interpreting timing measurements.

Timing measurements, especially periodic checks of opening and closing times performed on disconnected switching equipment are easy to carry out and very common. As a simple diagnostic test of the overall condition of the device, this is useful technique. However, if timing discrepancies are found, it may be necessary to apply more sophisticated diagnostic techniques, like travel curve measurements, to determine the root cause. Furthermore, normal operating times do not guarantee that a breaker is in good condition; several important failure types may develop without affecting the operating times of the unit.

Instruments for time measurements have for long been commercially available from several manufacturers of test equipment.

5.2.4.5   Pole Discrepancy

The term pole discrepancy usually refers to differences in operating times between the three phases or poles of a switching device within the same operation. Hence, in contrast to the operating time measurements discussed in the previous section, synchronous data collection from each pole is here required.

As stated earlier, timing information is often easy to obtain and provides a reasonably good estimate for the overall condition of a switching device. The same applies to timing discrepancies between the three phases of a circuit breaker.

Determining pole discrepancy in a circuit breaker on a continuous basis becomes less accurate because it in most cases has to be based on break time or make time recordings obtained through the current transformers. Break and make times vary somewhat with how the contact movement coincides with the current waveforms. Furthermore, at what time current zero occurs in the second and third pole to break, depends also on system earthing.

Circuit breakers where each pole has a separate operating mechanism are usually equipped with a system that trips the breaker if not all poles respond when a command signal for close is given. Such an incident can be considered as a rather extreme example of pole discrepancy (several hundred milliseconds). The system triggers only on unsuccessful closings; not on single-phase open-close sequences as in an automatic reclosure operation. The notification that follows to the system operator indicates that there is a severe problem with at least one of the poles.

Recording pole discrepancy is a simple and widely applied diagnostic technique, most commonly used on a periodic basis. It is considered a useful tool for all types of switching equipment, not only circuit breakers.

For circuit breakers, IEC states that the discrepancy in both opening and closing times should be less than one-half cycle of the rated frequency, unless otherwise specified by the manufacturer. Many manufacturers or utilities have stricter requirements, for example by setting a maximum acceptable deviation in opening times of 5 ms. Under some circumstances, e.g. after a single pole auto-reclosure where the charging of the operating mechanism may differ between the poles, these requirements are not applicable.

For most other types of switching equipment, pole discrepancy recordings must be evaluated with basis in knowledge and experience with the unit tested.

5.2.4.6   Arcing Contact Wear

The sum of the time integrals of the squared arcing current for opening and closing operations is used as a measure of the arcing contact wear, although the accumulated arc energy seem to be a better parameter for this purpose. This is mainly due to the difficulties on online arc voltage measurement. The inaccuracies in the determination of the arcing time, the DC offset that occurs under asymmetric faults, and variations of the arc voltage have however, to be taken into consideration. Hence, for very accurate measurements it is necessary to be able to precisely measure or calculate the waveform of the arc current and arc voltage.

Permanently installed systems for accumulating $i^2(t)$ over the arcing time t_{arc} have been on the market for several years, but not been applied in any great numbers. The reason is primarily economic; the equipment contributes significantly to the cost of the circuit breaker.

However, arcing contact wear is not a large problem on most modern circuit breakers. Very often, dismantling and inspection of the contacts of circuit breakers that have been in service for many years reveals that the arcing contacts are virtually as new.

An alternative method for detecting arcing wear on a periodic basis is dynamic contact measurements, i.e. by measuring the contact resistance using a low voltage direct current during a closing or opening operation. By using a rather large current, a high sensitivity in the resistance measurement can be obtained. For breakers with two contact pairs (arcing contacts and main contacts) the commutation of the current is seen as a shift in the resistance level, see Fig. 5.10. Excessive arcing contact wear can be detected as changes in the resistance characteristics.

Dynamic contact resistance measurements are increasingly applied as a periodic diagnostic test on circuit breakers, very often in combination with measurements of contact travel. The advantage is the simplicity of the method, and that detailed knowledge about the condition of the contacts is obtained.

Fig. 5.10 Two examples of dynamic contact resistance measurement from opening operations on the same type of circuit-breaker. In the *upper trace* the elapsed time between separation of main and arcing contacts is substantially shorter than the *lower trace*, indicating that the length of arcing contact rod has been significantly reduced, e.g. due to heavy arc erosion

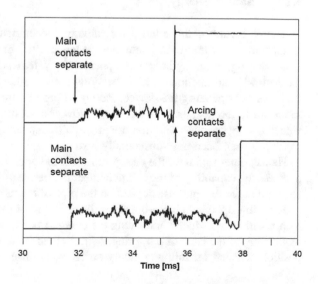

5.2.5   Mechanical Operation

5.2.5.1   Failure Mechanisms

The operating mechanism or the "drive" of a switching device provides the energy and the mechanical transfer needed to move the contacts from open to closed position and vice versa.

In circuit breakers, where a high contact speed is required, the operating mechanism generates and transfers large mechanical energies and forces. On most circuit breaker types, the transfer system also serves as electrical insulation between the moving contact and ground. Moreover, in many cases the mechanical linkage contains a large number of moving parts. Altogether, these factors make the drive among the more susceptible parts of a switching device.

Experience has shown that a large fraction of the failures on all types of switching equipment occur in the operating mechanism. For circuit breakers, the most common problems are oil and air leaks from hydraulic and pneumatic systems. Fortunately, these can normally be solved easily and without causing outages. However, more severe incidents such as mechanical breakdown of rods, shafts, springs, gears etc. during a switching operation do also occur, and the consequences can be dramatic.

Table 5.6 gives an overview of the diagnostic test methods and sensors relevant for the mechanical operation of the contacts of switching equipment. Several of the most important techniques, such as measurements of operating times and contact travel characteristics, are also applicable for testing of the "switching function", and have therefore, already been treated in Sect. 5.2.4.

5.2.5.2   Stored Energy

In circuit breakers, the energy for moving the contacts is usually supplied by springs, or by hydraulic or pneumatic systems, often in combination with springs. Leaks or clogging in hydraulic and pneumatic systems, as well as a wide variety of mechanical malfunctions and breakdowns may affect the mechanical energy available for operating the device. Failing to provide the correct amount of energy can result in too fast, or more likely, too slow contact movement. A reduced contact velocity is a serious condition for a circuit breaker as the current interruption capability then becomes substantially lowered.

Hence, determining of the energy stored in the operating mechanism can be used to assess the capability of the circuit breaker to operate as intended. The methods that are applied for this purpose depend on the type of operating mechanism considered.

In spring-operated circuit breakers, the charging of the spring can be determined with position transducers indicating the charged position directly or by continuous monitoring of the latch of the spring. As described in Sect. 5.2.4.3, a wide variety of position sensors, including auxiliary switches, are used.

Table 5.6 Diagnostic techniques and sensors for testing the mechanical operation of switching equipment

Parameter	Application(s)	Method/sensor	
Contact travel characteristics (position, velocity, acceleration)	All	Dynamic position sensor with analogue output:	
		Resistance potentiometer	C, P
		Magneto-resistive sensor	C, P
		Linear variable differential transformer	C, P
		Dynamic position sensor with digital output:	
		Optical with incremental coding	C, P
		Optical with non-incremental (absolute) coding	C, P
Operating time	All	Electrical recording of time to close/open of primary circuit	P
	ES, D	Motor running time	C, P
Pole discrepancy in operating times	All	Electrical recording of time to close/open of primary circuit	C, P
Stored energy	All spring operated equipment	Spring charging by:	
		Position transducer	C
		Charging time and/or current of motor	C, P
	All hydraulically operated equipment	Charging by:	
		Oil pressure	C
		Nitrogen pressure	C
		Position of pistons, springs, valves	C
	All pneumatically operated equipment	Air pressure in drive	C
State of motor	All equipment with motors	Motor current, voltage and temperature	C, P
		Running time per start	C
		Accumulated running time	C
		Number of starts	C
Number of operations	All	Counter	C
Vibration signatures	All	Accelerometers	C, P

Abbreviations
ES earthing switches
D disconnectors
C used for continuous monitoring
P used for periodic diagnostic testing

Normally, the springs are charged by an electric motor. The stored energy can thus be determined indirectly by measuring the motor current and voltage and/or the charging time. However, these methods are not very accurate, as they provide, to a large extent, information about the condition of the motor and the mechanical transfer between motor and spring, rather than the stored energy in the spring.

In circuit breakers with hydraulic operating mechanism, energy is stored by using a hydraulic system to pressurize nitrogen gas. Hence, the pressure in the oil or nitrogen is a measure of the amount of energy stored. Other options include position monitoring of pistons, springs and valve indicators as well as even complex schemes to determine the nitrogen density.

Similar approaches can be applied on pneumatic operating mechanisms. The pressure of the compressed air is by far the most commonly measured parameter.

Since the amount of energy available in the mechanical drive is a very important factor, all circuit breakers are equipped with systems that continuously monitor this parameter indirectly. If the pressure in the hydraulic or pneumatic operating systems is below a predetermined level or if springs are not sufficiently charged, the operator is alarmed and the breaker is blocked for operation.

However, the best method for assessing whether the drive provides the mechanical energy needed for correct contact movement is to measure exactly that parameter. Thus, recording the contact travel characteristics (see Sect. 5.2.4.3) is a very useful technique, also in this context.

Many circuit breaker types, especially models that are spring operated, are set to charge the drive only after closing operations. In the case of a fast open-close-open sequence, there will be no time for charging, so the energy of the drive will differ from what is seen after normal close operations. This must be taken into consideration when evaluating the measurements.

As mentioned above, oil and gas leaks in hydraulic and pneumatic drive systems are quite common. A simple and efficient method for monitoring the overall tightness of the systems is to count the number of starts or the total runtime of compressors and pumps within a given time period (day or week). These parameters are certainly also depending on whether the circuit breaker has been operated within the considered period, and this must be taken into account.

5.2.5.3 State of Motor

Electric motors are used in the mechanical drives of switching equipment mainly for two purposes: to charge the operating mechanism (springs or compressors) in circuit breakers, or for directly manoeuvring the contacts in earthing switches and disconnectors. Hence, the motors are normally only running in short intervals with a well-defined load, and are designed for this not very demanding duty.

However, a number of important failure mechanisms may drastically increase the strains on the motors, both in terms of power requirements and running time. For example, poorly working lubricants in mechanical linkages, contact assemblies or compressors will require a substantially higher torque. Leaks in the hydraulic system will cause a compressor motor to run more frequently and for longer times. Consequently, information about the condition of the mechanical drive can be obtained by considering parameters reflecting the state of the motor or the strains the motor has been subjected to.

These parameters include motor current and voltage, temperature and status of the protection circuitry. Furthermore, the number of motor starts can be monitored with a counter, and the accumulated running time with a timer. A particularly informative approach is to measure the running time per motor start. If this value increases significantly, this can be taken as a reliable indicator of problems with the device or system the motor is operating.

Using parameters obtained from the motor for monitoring the mechanical drive is quite common, although in some cases it can be difficult to distinguish drive problems from problems originating in the motor itself.

The simplest methods, such as reading off the accumulated running time for compressor motors on a periodic basis are mostly applied. More complicated schemes, such as continuous monitoring of the motor current, are less common.

5.2.5.4 Number of Operations

In general, wear and tear of mechanical systems increase with use. Thus, the number of operations a switching device has performed can be taken as a very general measure of the overall condition of the device, including the mechanical drive. Many operating cycles are assumed to be associated with greater wear and a higher failure rate. Consequently, in conventional maintenance philosophy the number of cycles is, with good justification, among the prime factors determining the maintenance intervals. All circuit breakers and a large fraction of disconnectors and earthing switches are equipped with counters that record the number of operating cycles.

However, many utilities have experienced that this simple relationship between number of operations and overall mechanical condition in many cases is not valid. The picture is somewhat more complicated because also circuit breakers that operate very rarely, like once or twice a year, show a higher probability for drive problems. Quite often the reason is that grease and other lubricants harden over time, primarily simply due to lack of "exercise".

Nevertheless, the number of operations is a very useful parameter for diagnostic purposes, especially when combined with knowledge and experience with the breaker considered. Many utilities take this parameter into account when assessing what type of diagnostic testing to carry out and when to do it, and also when planning invasive maintenance.

5.2.5.5 Vibration Signatures

Various types of mechanical malfunctions in the operating mechanism and in other parts of a breaker will influence the vibration patterns generated during switching

Fig. 5.11 Normal (phase S) and irregular (phase R) vibration signature from an opening operation of a 145 kV single pressure SF$_6$ breaker. Phase R was later dismantled and the deviation around 60 ms was found to be caused by an incorrectly adjusted moving contact

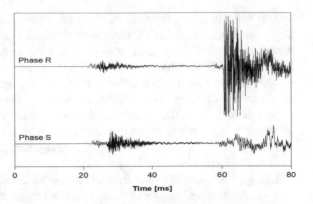

operations. Hence, by acquiring and comparing the vibration "fingerprints" or signatures with a reference, changes in the mechanical condition of the device can be detected as changes in the vibration patterns. The reference can be an earlier recording from the same breaker or the signature from another breaker of the same type.

Vibration patterns are obtained by using accelerometers and a standard data acquisition system. Typically, one accelerometer per operating mechanism is required. Vibration analysis can also be carried out on other parts of a circuit breaker, e.g. the arcing chamber of live tank circuit breakers. However, this requires that the unit is taken out of service and grounded; the method is thus limited to periodic diagnostic testing. Figure 5.11 shows an example on how a severe failure comes out as a significant change in the vibration pattern from an opening operation of a circuit breaker.

The major advantages with vibration analysis are that it focuses on mechanical type of failures, and that it is a general technique applicable on nearly all types of switching equipment. The main disadvantage is probably that rather sophisticated algorithms are required to produce a correct and quantitative measure on how similar or dissimilar two signatures are. Deviations have to be quantified to be able to distinguish between natural signature deviations and deviations caused by faults, with sufficient sensitivity and reliability.

Vibration analysis is an established diagnostic technique for rotating machinery, and has been introduced for switching equipment.

5.2.6 Control and Auxiliary Functions

5.2.6.1 Failure Mechanisms

Typically, a circuit-breaker operation is triggered by a 110–220 V DC command signal to a coil that moves a latch or opens a valve, which in turn releases the mechanical energy of the drive. For motor operated disconnectors or earthing switches, the command signal energizes a contactor that provides current to the motor.

Thus, the control system, comprising both the electrical circuitry and the latches and/or valves in the release mechanism, is a crucial part of the switching device. Reliability surveys have however shown that a large fraction of the failures occur in this part of the circuit breaker.

Failing to close or open on command is the predominant major failure type in the control circuitry, while delays in the operation is a typical example of a less severe, but rather common incident. A variety of underlying failure mechanisms exist, including low command voltage, defective coil and poor lubrication.

Table 5.7 summarizes the most important diagnostic techniques applied to the control system for switching devices. Included in the table are also a few techniques relevant for testing some of the auxiliary functions of a breaker.

Table 5.7 Diagnostic techniques and sensors for testing the control and auxiliary functions of switching equipment

Parameter	Application(s)	Method/sensor	
Coil current profile	CB	Recording of command current shape by:	
		Shunt	C, P
		Current transformer	C, P
		Rogowski coil	C, P
		Other current sensor types	C, P
Voltages	All	Voltage sensor	C, P
Status of auxiliary switches	All	Verification of integrity of device and control system based on:	
		Position	C, P
		Operating sequence	C, P
		Timing	C, P
		Mutual consistency of the various auxiliary switches	C, P
Circuit continuity	All	Continuous passing of small currents or current	C
		Pulses	
Environment in control cabinet	All	Temperature	C

Abbreviations
CB SF$_6$, air blast, minimum oil and oil tank circuit breakers
C used for continuous monitoring
P used for periodic diagnostic testing

5.2.6.2 Coil Current Profile

The conversion of the electric command signal to a mechanical signal, typically to move the tripping latch, turns out to be a very critical part of the operating sequence of a circuit breaker. Static loading of bearings and latches for long periods may impair the properties of the lubricants and lead to a large initial friction in the release mechanism.

Recording the profile of the current passing the release coil, as the command signal is applied, is a well-established diagnostic method for these parts of the control mechanism. Standard current measurement techniques such as shunts and Rogowski coils are applied. Fig. 5.12 shows a typical coil current signature. As indicated in the figure, the different parts of the process have their very characteristic features, and an experienced user can obtain an impressive amount of information about the condition of windings, lubrication, latches etc. from such traces. For example, the time from the coil armature starts moving until it stops, is a measure of how swift the release mechanism is operating, while the magnitude of the coil current at the plateau before the auxiliary contact opens gives the DC resistance of the coil. Short-circuits in the coil windings will reduce this resistance.

Figure 5.13 shows coil current signatures and operating times obtained from a minimum oil circuit breaker in service. For this circuit breaker model, the deviation in operating time between the phases should be less than 5 ms. This value is clearly exceeded is this case. Moreover, it is evident from the coil current measurements that the delays in phases S and T occur at around 10 ms, and that they most likely stem from a misadjusted or poorly lubricated tripping latch.

Examination of the coil current is a widely applied diagnostic method for testing the release mechanism of circuit breakers. It is efficient in detecting the most frequently occurring problems in the release mechanism and is rather simple in use. However, an accurate interpretation of coil current patterns requires experience and skills. A second important matter is that it focuses on only a limited part of the

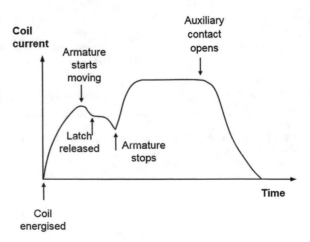

Fig. 5.12 Coil current signature (schematically)

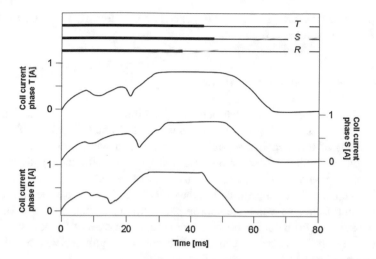

Fig. 5.13 Operating times (*thick/thin upper traces*) and coil current measurements from an opening operation of a 300 kV minimum oil circuit-breaker

circuit breaker assembly. Thus, many users prefer to employ coil current measurement in conjunction with other diagnostic methods.

Coil current measurements are primarily applied for periodic testing, and equipment for this purpose has for long been commercially available.

5.2.6.3 Control Voltage

Failures, e.g. open and short circuits in the control circuitry or in auxiliary parts of a switching device, may affect the voltage at the involved parts. For example, the voltage across a coil or a motor will change if some of the windings are short-circuited. Similarly, a defective heating element can be detected by considering the voltage drop across the element.

Thus, by measuring the voltage at certain locations, the control and auxiliary systems can be monitored in some detail. This is however, not very common as it requires a lot of wiring and a system for handling all the measurements.

It is more usual to monitor the voltage supplied to the control and auxiliary circuitry by measuring voltage at one point only. If this value is outside a specified range, this indicates problems with the circuitry or with the substation control system.

Incorrect supply voltage to a switching device is, strictly speaking, an indication of problems with the control system rather than the switching device itself, and thus beyond the scope of the present treatment. Still however, in the context of diagnostics it is important to bear in mind that an incorrect supply voltage in many ways may lead to incorrect diagnoses and false alarms. For example, motor running times and tripping coil current profiles will in general be significantly altered if the supply voltage is outside its specified range.

5.2.6.4   Status of Auxiliary Switches

Various failure types can be detected and identified simply by considering the auxiliary switches of the control circuits of a circuit breaker. The status (open/closed) of the switches as well as when and in what sequence their status changes during a circuit breaker operation, can be used for diagnostic purposes, both for the primary equipment and for the control system itself.

For example, a deviating operation time of a circuit breaker can be detected as a deviation in the time at which the appropriate auxiliary switch operates. Similarly, defective systems for blocking for breaker operation in case of too low gas density or insufficient spring charging can be identified by examining whether the right switches or contactors operate when they should.

Modern, advanced control systems for switching equipment facilitates an automatic and continuous monitoring of several aspects of the breaker operation based on the status of the auxiliary switches. For example, the sequence and timing of the operation of the various auxiliary contacts during a switching operation can be checked. Moreover, signals from several sensors can be crosschecked to verify that information conveyed from different sources is consistent.

For older equipment, the alternative is manual testing of the various systems. This can be performed as a periodic test, provided that the user has a detailed knowledge of the device. For performing diagnostic testing based on timing of the different auxiliary switches during a breaker operation, multi-channel testing equipment is commercially available.

5.2.6.5   Circuit Continuity

The continuity of an electrical circuit can be monitored by passing a small direct current or minute current through it. Interruption of this test current is a clear indication of irregularities, such as short-circuits and open-circuits.

Such monitoring systems are increasingly applied on important control circuits, e.g. the tripping system, of modern switching devices. The substation control system is automatically notified if defects are found, and the appropriate measures can be taken.

Permanently active heating elements are often installed in the control cabinet of outdoor switching devices. The circuit continuity can here be monitored by standard current measurement methods, like shunts, current transformers or Hall sensors.

These methods for monitoring the electric circuitry are efficient and well suited for continuous operation. They do, however, require comprehensive systems for acquisition, handling and interpretation of the measurements.

5.2.6.6   Environment in Control Cabinet

Switching equipment installed outdoors often have heating elements installed in the control cabinets to avoid very low temperatures in the control and operating mechanism during cold winter days, and also to avoid condensation of moisture in humid regions with rapid temperature changes.

A straightforward and direct method for surveying whether the heating elements and other parts of the temperature control system are working as intended, is to measure temperature inside the control cabinet with a standard sensor, e.g. a thermo-element. The sensor should be placed near the most critical parts of the control cabinet. Effects of sun radiation and ventilation should be considered when placing the temperature sensors.

References

1. For more information, visit the CIGRÉ webpage www.cigre.org
2. IEC Standard 62271. High voltage switchgear and controlgear
3. Cigré WG (1981) The first international enquiry on circuit-breaker failures and defects in service. Electra 79:21–91
4. CIGRÉ WG (1994) Final report of the second international enquiry on high voltage circuit-breaker failures and defects in service. CIGRE Techn Broch, vol 83
5. CIGRÉ WG (2012) Final report of 2004–2007 international enquiry on reliability of high voltage equipment, Part 2—Reliability of high voltage circuit breakers. CIGRE Techn Broch, vol 510
6. CIGRÉ WG (2000). User guide for the application of monitoring and diagnostic techniques for switching equipment for rated voltages of 72.5 kV and above. CIGRE Techn Broch, Vol 167

Chapter 6
Future Trends and Developments of Power Switching Devices

The previous chapters looked into various aspects of current interruption in conventional power grids including basic working principles and different applications of power switching components and technologies used in these devices. There has been a steady development of switching devices to increase their functionality, to add new features, to widen their application range and at the same time to reduce their production costs. One example is the investigations on application of vacuum switching technology for generator circuit breakers [1] and as high voltage circuit breakers [2, 3]. In addition, new developments and trends in power systems imply new functionalities and/or constraints on power components, in particular on switchgear. This chapter deals with some major upcoming issues in this regard: i) interruption of direct currents, ii) limitation of fault currents and iii) alternative gases as interrupting medium for new generation of high voltage circuit breakers.

The need for high availability and reliability of electricity supply leads to much more interconnections between different nodes of a power system, which in turn results in larger short circuit current levels. Very large fault currents are not only a challenging issue for power switches but also give high mechanical (and in some cases thermal) stresses on other components. One possibility to cope with the increasing short circuit current, is to design the power switching devices in such a way that they are able to reliably interrupt and handle very large short circuit currents. The other solution would be to have a new type of components, which shows a current dependent impedance (very low impedance by load current flow and very high impedance by short circuit current flow) and is able to limit the maximum prospective short circuit current levels of a network. Such devices are referred to as *fault current limiters*. If such components are available, high mechanical and to some extent also thermal stresses on other power grid components could be reduced drastically. Although, with the exception of fuses and fuse-like single-shot devices [4], no commercially available fault current limiter exists, but tremendous efforts have gone into investigate the possibility of realising such devices.

© Springer International Publishing AG 2017
K. Niayesh and M. Runde, *Power Switching Components*, Power Systems,
DOI 10.1007/978-3-319-51460-4_6

Another development in power systems is to use *High Voltage Direct Current* (*HVDC*) for transmission of electric power over long distances. It is expected that the increasing number of HVDC links will lead, in near future, to HVDC networks, i.e. interconnections of many HVDC power transmission links. Moreover, many renewable energy sources produce electric power as DC, several proposed energy storage systems are more compatible to a DC system, and in case of DC networks, there is no reactive power (e.g. to be provided to charge the capacitances of a long transmission system). Therefore, in many applications, use of DC may be a better solution than conventional AC systems. To be able to protect such HVDC networks, it is vital to interrupt the direct current faults, i.e. fault currents without a natural current zero crossing.

As mentioned in Chap. 4, SF_6 is the dominating interruption medium used in high voltage circuit breakers and many gas insulated substations and switchgears in high and medium voltage applications. This gas is unfortunately one of the strongest "greenhouse gases" and therefore, many investigations have been performed or are currently running to find more environmental friendly alternatives to SF_6 for power switchgear. The outcome of these studies may have a significant impact on the future trends and developments of gaseous power switching devices.

6.1 Interruption of Direct Current

6.1.1 Current and Voltage Waveforms

Interruption of direct current (DC) is fundamentally different from interruption of alternating currents as a DC current has no natural current zero crossing. This, among other things, leads to a more direct interaction between the arc and the system. This will be studied by using Fig. 6.1, showing a simplified circuit diagram for interruption in a DC system.

The breaker is open and an electric arc is burning between the contacts. Currents and voltages are determined by the circuit equation

Fig. 6.1 Circuit diagram for interruption in a DC circuit

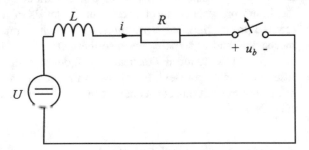

$$U = Ri + L\frac{di}{dt} + u_b \qquad (6.1)$$

This equation can be rewritten to:

$$\frac{di}{dt} = \frac{1}{L}[U - Ri - u_b] \qquad (6.2)$$

At the instant of contact separation, the entire voltage drop in the circuit was across the resistance R (There is no voltage drop across an inductance carrying DC). $U - Ri$ was therefore equal to zero. Consequently, right after contact separation

$$\frac{di}{dt} = -\frac{u_b}{L} \qquad (6.3)$$

As the burning arc will cause a voltage drop, u_b, across the switchgear, the time derivative of the current is negative. Consequently, the current i in the circuit starts to decrease when the contacts are separated. To be able to interrupt the current successfully, it must decrease to zero. Whether this happens or not cannot be determined from these expressions alone.

For example, assume the contacts have fully separated without the current being interrupted. The arc is now burning under approximately stationary conditions (no forced cooling, no contact movement, etc.). The arc voltage is then determined by the static arc characteristic (see Fig. 2.8) valid for the particular arcing medium and contact configuration. At static/stationary conditions $di/dt = 0$, so (6.1) becomes:

$$U - Ri = u_b(i) \qquad (6.4)$$

The left hand side of (6.4) is a straight line defined by the circuit parameters, whereas the right hand side is the static arc characteristic. The current–voltage relationships of each side of (6.4) are shown graphically in Fig. 6.2.

Figure 6.2 shows that (6.4) is fulfilled in the two points where the two curves intercept. The current was equal to U/R prior to opening the breaker, and as described above, the current starts decreasing when the contacts separate. The right

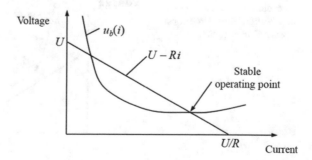

Fig. 6.2 Voltage across the contacts as a function of current when interrupting direct current

point of intersection is first reached. It can be seen that this point is stable by considering the effect of a small disturbance. Assume that the current for some reason, becomes slightly smaller than the value in the point of intersection. Then, $U - Ri$ (the straight line) minus u_b (the arc characteristic) is positive. By inserting $U - Ri - u_b > 0$ into (6.2), the result is $di/dt > 0$, i.e. the current increases. In the same manner, it can be shown that a disturbance giving a current that is slightly greater than in the point of intersection, $di/dt < 0$ and the current decreases towards the value of the point of intersection.

Therefore, in the right point of intersection $di/dt = 0$, the arc continues to burn, and there is no current interruption. Within a short period of time, the energy dissipation due to the arc becomes so large that the switchgear is destroyed (explodes).

Thus, to have a successful interruption there cannot be any stable point of operation, i.e. there should not be any points of intersection between the straight line and the static arc characteristic. The circuit parameters and the arc properties must be as shown in Fig. 6.3. Here, di/dt is negative for all i; the current goes to zero and is interrupted.

Note that when the current falls below a certain value, the arc voltage must exceed the system voltage. This is possible because the static arc voltage is not dependent on the system voltage, but is solely a function of the current flowing in the circuit.

The description is simplified, as it is based on the static arc characteristic. In reality the arc is dynamic, both because the arc length varies during the interruption, and the arc has a certain thermal inertia. These and other factors must be included in a more precise description. However, the crucial criterion remains; an arc voltage greater than the system voltage is required to create a current zero and thus a successful interruption of a direct current.

There are several ways of achieving this. As mentioned in Chap. 2, most of the voltage drop in an arc is generated close to the electrodes; the arc length is in this respect less important. A common way of increasing the voltage across a breaker is to split one arc into several short ones connected in series. This increases the number of anode and cathode voltage drops and a larger total voltage drop is thus

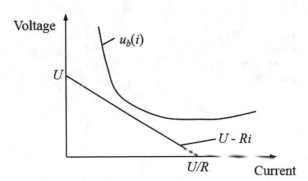

Fig. 6.3 Interruption of direct current. The arc voltage drop is sufficiently large for a successful interruption

achieved. Although this method has been successfully applied for interruption of direct currents in power networks with rather low nominal voltages, it is not directly applicable to high voltage direct current interruption.

6.1.2   Energy Dissipation

The energy dissipation in DC switchgears is a very important quantity. This will be studied by considering a simple circuit shown in Fig. 6.4, containing an inductance L on the system side, a switchgear and a resistive load. The contacts open, an electric arc is established and there is a voltage drop u_b across the breaker. The voltage across the load is u_l.

The circuit equation is:

$$U - L\frac{di}{dt} - u_b - u_l = 0 \tag{6.5}$$

If the contacts separate at time $t = 0$, and the current is interrupted at $t = t_1$, the energy dissipated in the breaker during the interruption is given as:

$$W_b = \int_0^{t_1} i(t)\, u_b(t)\, dt \tag{6.6}$$

By inserting (6.5) into (6.6), the following expression is found:

$$W_b = \int_0^{t_1} \left(U - L\frac{di}{dt} - u_l \right) i\, dt \tag{6.7}$$

Equation (6.7) can be transformed to:

$$W_b = \int_{i_0}^0 -Li\, di + \int_0^{t_1} (U - u_l)\, i\, dt \tag{6.8}$$

Fig. 6.4 Interruption of a load current in a DC circuit

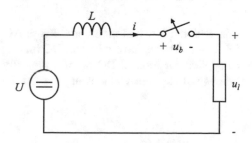

where i_0 is the load current. Integrating the first term gives for the energy dissipated during an interruption:

$$W_b = \frac{1}{2}Li_0^2 + \int\limits_0^{t_1} (U - u_l)\,i\,dt \qquad (6.9)$$

The second term in the right hand side of (6.9) is always positive, as the system voltage is always greater than the load voltage. This implies that the energy dissipated in the switchgear is equal to or greater than the magnetic energy stored in the circuit prior to the interruption. An additional contribution to the total energy dissipation depends on the load and the interrupting time.

6.1.3   Fault Clearing Time Requirement

Another important difference between the AC and DC systems that has great implications on switching device requirements is the time constant of the system as such and the fault current characteristics.

The contribution from transformer windings causes the overall inductance in an AC system to become substantially larger than in a DC system (which is without transformers). A consequence of this is that the overall time constant becomes smaller in DC systems. In AC systems, short circuit current clearing times of up to several power cycles or in the range of hundreds of milliseconds are acceptable. A short circuit in a DC system may have fault current rise times of only a few milliseconds, and the entire system may experience a massive voltage collapse soon after. Hence, in general, DC breakers set to clear faults must react much faster than AC breakers with a corresponding duty. For large HVDC system, short-circuit clearing times must be no more than 5–10 ms. Bearing in mind that modern transmission level voltage AC circuit-breakers require at least around 20 ms only to mechanically open the contacts, it is obvious that the reaction time requirement contributes to making DC interruption at the same voltage complicated.

6.1.4   DC Switching Devices

Conventional low and medium voltages AC switchgears, such as vacuum, air and SF$_6$ based devices have a certain DC interrupting capability. Hence, these can also be applied for modest DC ratings, typically seen in traction and industrial applications with system voltages up to a few kilovolts.

The fast reaction time and the large amounts of energy dissipated remain the main obstacles against the development of DC switchgears for higher ratings. For transmission system levels (hundreds of kilovolts), no DC device that is able to interrupt load currents or even more so, short-circuit currents (as the AC circuit-breaker does in an AC system) is at present commercially available.

This has great implications and is the main reason why branched DC transmission grids cannot be built. Since circuit-breakers are not available, selective protection schemes and disconnection of faulty parts of the system, as is done in AC systems, is impossible. Consequently, existing HVDC transmissions are point-to-point links, situated in a surrounding AC system.

Moreover, the topology of HVDC substations becomes very different from those of AC systems. A prominent feature is that instead of using switchgears that *interrupt* the current, devices that *commutate* currents are extensively used. For example, if a converter unit of a HVDC substation experiences a failure, the current into this unit is not interrupted, but instead commutated to a parallel branch that bypasses the failed components.

Figure 6.5 illustrates the difference between the current interruption and the current commutation functions. Important in this context is that current commutation is a far less onerous duty than interruption. The voltage drop across the switchgear and the energy dissipation become much lower when redirecting a current flow from one branch to another than when interrupting it.

In cases where a short-circuit current needs to be interrupted because the failed component cannot be bypassed, this is done on the AC side of the converters by using AC circuit breakers.

The various types of commutation or bypass switches used in HVDC substations are typically modified AC circuit-breakers, often with SF_6 as the interrupting medium. Ratings and testing requirements of such switchgears are not yet covered by the standards, and must thus be agreed upon between the equipment supplier and the customer for each project. Hence, in contrast to AC switchgear, switching equipment for transmission level DC substations are largely based on tailor-made solutions and case-to-case arrangements.

Although interruption of faults on the AC side and commutation of current from faulty parts may be applicable for point-to-point DC systems, these are not suitable solutions to protect the multi-terminal DC networks in case of short circuit faults.

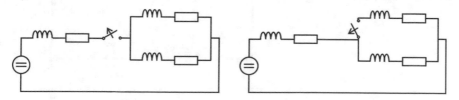

Fig. 6.5 Schematic DC circuits with switches performing a current interrupting (*left*) and a current commutation from one circuit to an existing parallel circuit (*right*). For higher system voltages and large current ratings only the latter type of devices are commercially available

As described in Sect. 6.1.1, a current zero crossing must be artificially created to be able to interrupt the current in a DC system.

6.1.5 Methods to Create an Artificial Current Zero Crossing

One principle used in DC switchgear to create a current zero is to inject or superimpose a current of opposite direction to the current to be interrupted. The superimposed current can be provided by interaction with a parallel oscillatory circuit or by actively injecting a current. In this way, an artificial current zero crossing is generated. The interruption is then rather similar to that of an AC circuit. The disadvantage of these methods is that the auxiliary circuit becomes large and expensive, as it must supply both large currents and high voltages.

Figure 6.6 shows the typical passive oscillatory circuit consisting of an inductor in series with a capacitor. To interrupt the current, the switch opens and an arc ignites between the contacts. The generated arc voltage increases in amplitude as current is reduced, see the arc characteristic of Fig. 2.8. Under certain conditions the current–voltage characteristic of the arc causes an unstable resonance LC-circuit to be established. This is because the time derivative of the current is negative, and the amplitude of the current oscillation increases. After a few oscillations, the amplitude of the resonance current is such that a zero crossing occurs in the arcing chamber, and the current is interrupted.

The current of the main circuit now commutates to the parallel branch and the capacitor starts to charge. The surge arrestor in the second parallel branch limits the generated voltage, which opposes and exceeds the system voltage. It also absorbs the inductive energy of the system, forcing the current in the system to zero.

Figure 6.7 shows a scheme for actively injecting a current in a direction opposite to the system current and thereby creating a current zero. While the switch in the main circuit is closed, a capacitor is pre-charged from a separate charging unit. The interruption process starts with first opening the contacts of the switch in the main

Fig. 6.6 DC breaker with a parallel oscillation circuit and a parallel surge arrestor

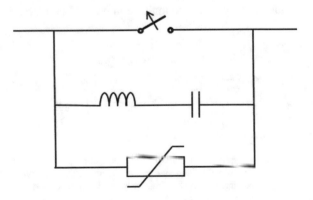

Fig. 6.7 DC breaker with active current injection from a pre-charged parallel capacitor

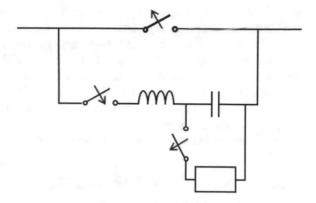

circuit followed by closing the making switch in the parallel circuit. This causes the capacitor discharge current to be superimposed on the current in the main circuit. If the discharge current amplitude exceeds that of the current in the main circuit, an artificial current zero is created, and current is interrupted.

As the maximum reaction time of a DC interrupter must be few milliseconds, the opening mechanical switch must be very fast. It is possible to use solid-state switching devices in the main path, but this results in very high losses during the nominal current flow and leads to very expensive, complicated solutions.

As it can be seen from the descriptions above, the task of current interruption in DC systems is to a great extent based on current limitation or suppression. Hence, the topic of current limitation, which will be dealt with in the next sub-chapter is highly relevant for DC switching devices.

6.2   Fault Current Limitation

Although conventional power switching devices have a dynamic impedance characteristic initiated by opening or closing operations, the voltage drop caused by the arc is almost negligible compared to the network voltage. Therefore, the current flowing through a power switch experiences almost no change as its contacts separate and an arc ignites. Thus, fault current limitation is not inherently integrated in the functionality of switchgears, apart from certain low voltage current limiting circuit breakers and fuses.

Insertion of passive constant impedances such as air cored reactors in the current path is a way to reduce the short circuit level of a network. The disadvantage is the substantial continuous losses, which impose limitations on energy supply and voltage regulation of the loads, especially in case of inductive loads.

The practical implementation of a constant impedance is in many cases in form of a current limiting reactor, which could be air-cored or with a gapped magnetic core to avoid saturation at high currents. Application of such a current limiting reactor may increase the system inductance, reduce the power factor and increase the rate of rise of the transient recovery voltages seen by the switching devices.

The following simple example shows how the constraint of having a minimum acceptable load voltage limits the applicability of a constant current limiting impedance. For this purpose, assume that a minimum 95% of the rated voltage is desirable at the load. The short circuit impedance of the system has been considered as a short circuit inductance and a current limiting reactor with a total inductance of $L_{limiter}$ is used (see Fig. 6.8).

In this simplified circuit, the load voltage is given as:

$$U_{load} = U_n \frac{Z_{load}}{Z_{load} + j\omega(L_{sc} + L_{\text{limiter}})} \qquad (6.10)$$

To fulfil the minimum load voltage requirement of $|U_{load}| \geq 0.95|U_n|$, the following inequality has to be satisfied:

$$|Z_{load}| \geq 0.95 |Z_{load} + j\omega(L_{sc} + L_{\text{limiter}})| \qquad (6.11)$$

In the worst case, when the load is inductive, this results to a minimum limited short circuit current of 19 fold of the rated load current.

Another more ingenious way would be to have components with current dependent impedances. In an ideal case, such a component has to act as a short circuit when the load currents flow, and show very large impedances under fault conditions, i.e. when the short circuit current flows. As this impedance exists only under fault conditions, the load flow and voltage regulation are not affected by such kind of fault current limiters.

Different non-linear phenomena may be used to realize a non-linear current controlled impedance. Different non-linear material properties like change of resistivity with current or temperature and saturation of ferromagnetic materials are utilized. In this type of fault current limiters, the non-linear impedance is supposed to be able to carry the nominal load current and therefore the voltage drop over this

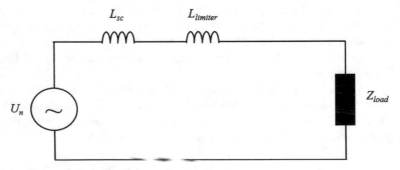

Fig. 6.8 Simple circuit to describe the limitation on constant current limiting impedance (reactance)

should be in the acceptable range of a few per cents of the rated voltage of the network. Thus, a limitation on the maximum acceptable impedance value for the fault current limiter under normal conditions is applied. Due to the limited dynamic range of material properties, this may result in insufficient limiting impedances. To explain this limitation, consider a non-linear impedance Z, where the ratio of its final value Z_f to its initial value Z_0 is equal to k (i.e. $k = \frac{Z_f}{Z_0}$). Furthermore, the maximum voltage drop on the fault current limiter under normal working conditions is assumed to be $\alpha\, U_n$, where U_n is the rated voltage of the network. It can be easily shown that the limited short circuit current has a minimum:

$$I_{SC}^{\text{limited}} \geq \frac{I_n}{\alpha \cdot k} \tag{6.12}$$

where I_n is the rated current of the network. It must be noted that (6.12) is derived based on the assumption that the short circuit impedance of the network is supposed to be much smaller than the fault current limiter impedance and for the sake of simplicity has been neglected.

Considering (6.12), the limited short circuit current is very dependent on the dynamic range of the fault current limiter impedance as well as on the maximum tolerable voltage drop.

The limited dynamic range of the fault current limiter impedance and the thermal issues related to the losses associated with continuous current flow through the non-linear impedance make the fault current limiters of this type very large and expensive.

Another approach is to combine non-linear material properties with dynamic circuit topology changes, e.g. using fast opening switches in parallel to non-linear current controlled impedances. This method has the advantage of not having a continuous current flow through the non-linear impedance. The consequence would be a much simpler and compacter design of the non-linear impedance.

In this type of fault current limiters, not the non-linear impedance itself, but a switch placed in parallel to the non-linear impedance, carries the load current and only by occurrence of a fault, the current is commutated to the impedance and is limited. The drawback of this method is that the limitation of the current does not happen inherently and therefore, a control unit is necessary to detect the fault occurrence and to send the opening command to switch.

A comprehensive survey on different methods studied to realize fault current limiters may be found in [5]. In the following, some of the most promising techniques are briefly explained.

6.2.1   Superconducting Fault Current Limiters

One of the non-linear current-controlled phenomena, which can be used for current limitation purposes, is superconductivity. In superconducting materials, a rapid resistance change may be realized when the current or magnetic field exceeds a threshold that causes the superconductor to quench, i.e., leaving the superconducting state and becoming resistive. This transition phenomenon may be used in different ways to limit fault currents.

One possibility is to use the superconducting device directly in the current path. In this way, a non-linear current-controlled resistor represents the superconducting fault current limiter (see Fig. 6.9). A parallel shunt resistor is used to limit the voltage drop across the superconducting device during fault current limitation. In this way, the superconductor itself acts as a switch, which commutates the current to the parallel shunt resistor. The series circuit breaker interrupts the limited short circuit current and isolates the faulty part of the network. The fault current limitation process in the superconducting material produces heat, which needs to be removed before this device is cooled down and put back into service again.

Another approach is to place the superconducting device in parallel to the main circuit. One realization is to use a transformer like two winding iron cored mutual inductance and to short circuit the secondary winding as shown in Fig. 6.10. This concept is referred to as the shielded core superconducting fault current limiter. Under normal conditions, as the superconducting device has zero resistance, the impedance seen from the primary side is very small (about as high as the stray inductance of the primary winding). Under fault conditions, flow of higher currents in the secondary winding through the superconductor results in a rapid increase of its resistance. As the consequence, the short-circuited transformer is changed in a very short time to an open-circuited transformer. Therefore, the impedance of the fault current limiter seen from the power network changes from stray inductance of the primary winding (very low value) to the magnetization impedance of the

Fig. 6.9 Resistive type superconducting fault current limiter

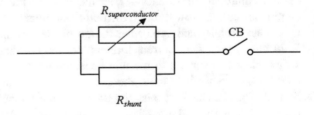

Fig. 6.10 Schematics of an inductive (shielded core) superconducting fault current limiter

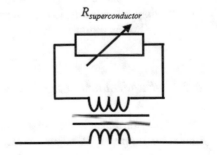

transformer (very high). A non-linear current-controlled inductance represents this type of superconducting fault current limiter.

Unfortunately, there is no material with superconducting properties at room temperatures. The ordinary or metallic superconducting materials show this property at temperatures lower than 30 K. The so-called high temperature superconductors (HTS) have the transition to superconducting phase at temperatures far below 200 K. Therefore, a cryogenic system (e.g. based on liquid nitrogen) must be used to cool down the materials to the desired temperatures. This makes fault current limiters based on superconducting materials rather complex and expensive. There are many active groups around the world working on different designs of fault current limiters based on superconducting materials, where both resistive and shielded-core approaches are used [6, 7]. Some prototypes are constructed and installed in some grids to get operational experience with these types of fault current limiters. However, the complexity and high cost of superconducting fault current limiters remain main challenging issues to introduce them as commercial products.

6.2.2   Saturable Core Fault Current Limiters

Another non-linear phenomenon, which may be used for current limitation purposes, is the saturation of ferromagnetic materials. This phenomenon is associated with change of relative permeability of ferromagnetic materials from very small values near unity to very large values of thousands or tens of thousands. A typical scheme may consist of a biasing magnetic field generation system (e.g. using a DC current coil [8] or permanent magnets [9]) to bring the core to saturation under normal working conditions of the fault current limiter, and at least two AC coils, which are normally connected in series and put in the main current path (Fig. 6.11). Biasing magnetic field is selected in such a way that the core remains in saturation, when the load current flows through the AC coils. In case of a fault, the increased current flowing through the AC coils produces large enough magnetic fields to bring the core out of saturation. This is observed as an impedance increase of the AC coil, as the permeability of the ferromagnetic core is increased dramatically.

It is also possible to use the superconducting coil in this design, which is then known as a saturable coil superconducting fault current limiter [10]. The superconductor coil is used solely to reduce the losses related to the continuous current flow through the DC coil; this means, unlike the resistive or inductive superconducting fault current limiters, the transition from superconductor to normal conductor as used in resistive and shielded core superconducting fault current limiters, is not applied here.

Although there are a number of field installations of such saturable core fault current limiters to reduce the maximum prospective short circuit current level of power networks, this type of fault current limiter has not become a commercial product yet, due to its very large size, weight and high cost.

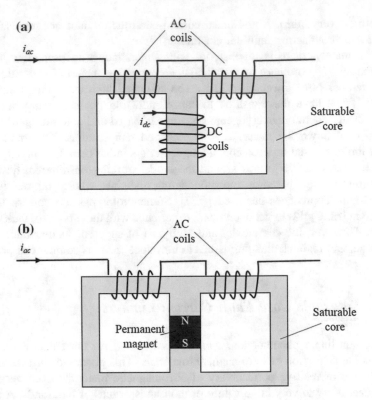

Fig. 6.11 Saturable core fault current limiter with the biasing magnetic field based on **a** DC current coil **b** permanent magnet

6.2.3 Solid-State Fault Current Limiters

In this class of fault current limiters, either solid-state switches are used to initiate a topology change in a circuit and commutate the current to other parallel paths containing current limiting or high impedance components, or their non-linear behaviour is used to design current-controlled impedances. There are many different approaches, but all designs suffer from high losses in nominal current path. The interested reader is referred to the interesting review paper on different types of solid-state fault current limiters [11].

6.2.4 Hybrid Fault Current Limiters

In the discussed methods for current limitation, the current limiter is put in the nominal current path. Therefore, some contradictory constraints have to be fulfilled, and many issues regarding the losses due to the continuous current flow through the fault current limiter have to be managed. This makes the designs of such devices complex and results in large and expensive solutions. One idea would be to separate the current carrying function of fault current limiters from their current limiting function during a fault occurrence. Such an idea may be realized if a very fast opening mechanical switch is used to carry the nominal load current under normal conditions. In case of a fault, the mechanical switch is opened and the current is commutated to one or more parallel paths. With this idea, all possible fault current limiter schemes may be used in the parallel branch. This approach is schematically shown in Fig. 6.12.

As seen in Fig. 6.12, a hybrid fault current limiter has three major sub-components, namely a fast opening mechanical switch, a measurement and control unit and a current limiting component or circuit. The fast opening switch has to be designed to open with very short opening delay times and at the same time to generate an arcing voltage, which is high enough to commutate the current from this switch to the actual current limiting component. The measurement and control unit is to decide on opening of the fast mechanical switch based on the simultaneous value of current, the rate of rise of current or a combination of both.

The drive mechanism of conventional switching devices is too slow to be used in such a fast mechanical switch. There have been many ideas on how to realize the necessary short delay times using the so-called electrodynamic drive mechanism [12]. In this approach, a pulsed current flows through a coil by discharging of a pre-charged capacitor. In the metallic disc adjacent to the coil, an eddy current is induced due to the very fast variations of the produced magnetic field, which in turn produces another magnetic field. Interaction of these two magnetic fields (primary magnetic field of the coil and the magnetic field produced by the induced current) results in a very large repulsion force, which could be used to accelerate the moving part of a fast opening mechanical switch.

In hybrid fault current limiters, the voltage drop during the normal operation causes no limitation anymore and therefore the initial impedance of the fault current limiter can be much larger compared to the situation, where the fault current limiter is put in the main current path. The challenge here is the current commutation from the fast mechanical opening switch to the current limiting component. In some cases to facilitate the current commutation process, a semiconductor switch is used

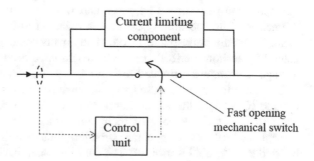

Fig. 6.12 Schematics of a hybrid fault current limiter

in series to the fast opening mechanical switch [13]. This is a trade-off to reduce the losses during the normal operation of the device but at the same time to increase the initial impedance of the current limiting component, and to have a simpler design of the fast opening switch.

Furthermore, for the thermal design of the current limiting part, unlike the FCLs directly placed into the main current path, the normal operation losses are of no concern. The important quantity for dimensioning of this part is the maximum energy, which has to be absorbed by the limitation of the current. Like as the case of HVDC circuit breaker [see Eq. (6.9)], this energy is greater than the magnetic energy stored in the short circuit inductance of the network at the time of current limitation.

6.3   SF$_6$ Free Power Switching Devices

As described in Chap. 4, the excellent dielectric and arc quenching properties have made SF$_6$ the dominating interrupting medium for circuit-breakers at transmission level voltages. Also in switching devices at lower voltage levels and with lower current ratings SF$_6$ has been used extensively. However, around 1990 it became clear that certain physical properties of SF$_6$ makes it a very potent "greenhouse gas", so when released to the atmosphere it contributes to "global warming". As a consequence, the industry started looking for more environmentally benign alternatives. This search for new gases for power switches intensified as some governmental bodies and regulators after 2000 started discussing taxing and even banning SF$_6$.

After briefly reviewing the properties of SF$_6$ that triggered these efforts, some of the alternative gases and technical solutions under consideration are presented. The main source of information is work reported in CIGRÉ and CIRED [14–18].

6.3.1   Environmental Concerns Related to Emissions of SF$_6$

The surface of the earth emits heat radiation. A part of this radiation does not go into space but is absorbed by gases and water vapour in the earth's atmosphere. A fraction of this absorbed heat causes radiation to be re-emitted and returned to earth. Thus, the atmosphere in effect thermally insulates the earth. Without this natural "greenhouse effect", the earth would have been a very cold place, with an average temperature estimated to around -19 °C, rather than the present 15 °C.

Human activity has released vast amounts of various gases, in particular carbon dioxide (CO_2) into the atmosphere. Some of these gases absorb heat to such an extent that further increase of their concentration in the atmosphere causes the "greenhouse effect" to become stronger, thus contributing to what is usually referred to as "global warming". The contribution of each gas type is determined by several factors such as the physical properties of the gas, the amounts emitted, and its atmospheric lifetime. In addition to water vapour and CO_2, the main contributors to global warming are methane (CH_4), ozone (O_3) and nitrous oxide (N_2O).

The amount of SF_6 released to the atmosphere is minute compared to CO_2 and CH_4, and so is its contribution to global warming. However, SF_6 is an extremely strong greenhouse gas, primarily due to its strong heat absorption and its long-term stability. The lifetime of SF_6 in the atmosphere is more than 3000 years, whether for example methane disintegrates within 12 years. Compared with CO_2, SF_6 is around a 23,000 times stronger greenhouse gas. In other words, emission of 1 kg SF_6 to the atmosphere has the same negative effect on the climate as releasing 23 tons of CO_2. Therefore, limiting the already small emissions of SF_6 is cost-effective compared to many of the other measures considered for limiting global warming.

6.3.2   Alternative Gases

In the following sections, a few results from investigations of alternative gases are reviewed. Dielectric insulating performance and current interrupting performance are assessed and comparted with SF_6. In addition, other properties such as toxicity, degradation and impact on climate changes are considered.

6.3.2.1   Carbon Dioxide (CO_2)

The capabilities of several common gases such as air, nitrogen and hydrogen were dealt with in Chap. 4. These are all fairly good choices for use in switchgear, but clearly inferior to SF_6.

Carbon dioxide, which is also a common and readily available gas, shows about the same dielectric strength as air and nitrogen, i.e. one third of SF_6 at atmospheric pressure. For higher pressures, the difference is somewhat less; at the pressure of 6 bar, CO_2 is found to have about 40% of the dielectric strength of SF_6. Hence, a dielectric design based on using CO_2 at a moderately higher pressure than the typical 6 bar currently used in SF_6 equipment emerges as an interesting option. For example, 10 bar CO_2 has about 70% of the AC and lightning impulse withstand levels of SF_6 at 6 bar.

With regard to the current interrupting capabilities of CO_2, the situation is quite similar. At 10 bar a CO_2 breaker is able to interrupt currents of some 2/3 of the magnitude of that of SF_6 at 6 bar. Adding 10–20% of oxygen to the CO_2 has been found to improve the interrupting capability somewhat.

As a rough overall comparison between SF_6 and CO_2-based circuit breakers, it has been suggested that a 10 bar CO_2 breaker can meet the requirements one voltage and current class below that of a 6 bar SF_6 breaker. Hence, a 6 bar SF_6 circuit-breaker rated for 550 kV/40 kA may be able to comply with the

420 kV/31.5 kA class test requirements if slightly modified and instead filled with 10 bar CO_2.

CO_2 recombines after arcing, so—similarly to SF_6—there is no need for topping up with new gas after many interruptions. CO_2 is non-toxic, but when dissociated by the arc and exposed to the materials inside the arcing chamber many new substances form. Although probably far less hazardous that some of the SF_6 arcing by-products, it may be wise to exercise some caution when dealing with the gas and the internal parts of arcing chambers that have experienced many interruptions of large currents.

The global warming potential (GWP) value for CO_2 is (by definition) equal to unity, so emissions to the atmosphere of the amounts used in switching equipment are not harmful. Moreover, CO_2 at 10 bar can cover all low temperatures requirements (down to −50 °C) without the problems caused by liquefaction of the gas. This is a major advantage with CO_2 as it makes this gas well suited for use in outdoor circuit breakers in very cold climates.

The first commercial high voltage circuit-breaker product with CO_2 as the arcing medium came in 2012, see Fig. 6.13. It is rated for 72.5 kV/2750 A/31.5 kA and has a 10 bar filling pressure. The arcing chamber, contacts and nozzles appear to be modified versions of standard SF_6 technology.

6.3.2.2   Fluorketone ($C_5F_{10}O$) and Fluornitrile (C_4F_7N)

Systematic screenings and searches for alternative gases identified several synthetic fluorine-containing gases as potentially interesting for switchgear applications. More examinations and tests reduced the number further, so in particular two gases stand out as promising candidates: a certain fluoroketone and a certain fluoronitrile. Figure 6.14 shows their molecular structures. They both have a carbon backbone, with five and four atoms, respectively.

Both the fluoroketone and the fluoronitrile have a substantially higher dielectric strength than SF_6, about a factor two at the same pressure. However, the boiling point is significantly higher than for SF_6. For pure fluoroketone, the boiling point at atmospheric pressure is 27 °C and for fluoronitrile it is −5 °C. The important practical consequence is that these gases must be mixed with a buffer or carrier gas to be able to use them at low temperatures; otherwise, they condense to the liquid state with a very low vapour pressure and thus show a poor dielectric strength and poor current interrupting performance. Several carrier gases have been proposed and tested, such as air, nitrogen, oxygen and carbon dioxide.

Finding an appropriate concentration of fluoroketone or fluoronitrile in the carrier gas, as well as the total gas pressure becomes a trade-off between the dielectric strength obtained and usable temperature range. A high partial pressure of fluoroketone or fluoronitrile and a high total gas pressure give a sufficient dielectric strength, but cannot cover the entire temperature range usually specified for outdoor circuit breakers.

Fig. 6.13 Circuit-breaker filled with 10 bar CO$_2$ and rated for 72.5 kV. In this pilot installation, it operates in the 145 kV system (Courtesy of ABB)

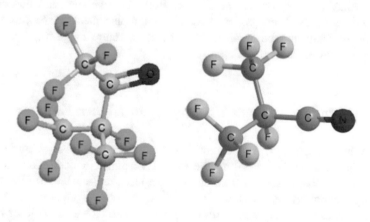

Fig. 6.14 The molecules of the fluoroketone (*left*) and the fluoronitrile (*right*) proposed as an alternative to SF$_6$ in switchgear

For example, a mixture of 6% fluoroketone, 81% CO$_2$ and 13% O$_2$ has at the pressure of 7 bar about 90% of the dielectric strength of 4 bar SF$_6$. The switching performance is also somewhat lower than for SF$_6$, but more importantly, the lower temperature limit is 5 °C. Hence, this is only for switching equipment installed indoor, whereas conventional SF_6 circuit breakers typically operate down to −40 °C.

Gas mixtures of fluoronitrile and CO_2 are usable for a considerably wider temperature range. At −25 °C and around 8 bar the dielectric strength and current interrupting performance are only slightly below those of SF_6 at 5.5 bar. However, it is clear that transmission level circuit breakers rated for high short circuit currents and −40 °C ambient cannot be based on these gases.

The fluoroketone mixtures have a GWP value of unity, and is thus from climate perspective a good choice. Pure fluoronitrile has a GWP-value of 2100, which is clear disadvantage for a gas proposed as an environmental friendly alternative to SF_6. Using a gas mixture that only contains some 5–10% fluoronitrile, the GWP value lowers correspondingly, but it remains far from negligible.

As opposed to SF_6, neither fluoroketone, nor fluoronitrile recombine after being exposed to an arc. Thus, to a certain extent these gases are "consumed" during switching operations. Estimates of the amounts decomposing suggest that this is not critical in the long term, at least not in breakers with arcing chambers of normal volumes subjected to normal test duties.

Preliminary investigations of toxicity indicate that certain harmful arcing by-products are formed, and it is recommended to treat these gases when they have been exposed to arcing in the same way as SF_6. More work is required to clarify all aspects of toxicity and also their compatibility with the other materials used in switchgears.

For distribution level switchgears (typically 6–36 kV), conventional SF_6 insulated equipment have much lower gas pressures that transmission level circuit breakers; they typically operate at 1.3 bar. Both fluoroketones and fluoronitriles mixed with air have been proposed as a direct replacement for SF_6 as electric insulation medium in such applications. The temperature range required is here normally not lower than −15 °C.

6.3.3 Extending Vacuum Technology to Higher Voltages

As described in Chap. 4, the current interrupting capability of vacuum circuit breakers is excellent, so extending the use of vacuum circuit breakers to transmission level voltages (several hundred kilovolts) as a replacement of SF_6, vacuum switching technology seems at first sight to be an obvious alternative. Vacuum switchgear does not in any way harm the environment. However, a closer look on how the different breaker technologies scale in terms of voltage ratings, reveals that such an approach is flawed, see Fig. 4.12.

The rated voltage of switchgears based on SF_6 and other gases can be increased fairly easily by increasing the physical dimensions of the device and the gas pressures. To the first approximation, the dielectric strength of a gas gap is proportional to the gap distance and to the gas pressure. Moreover, the thermal interruption performance of a breaker can be increased by a more powerful gas blast on the arc and a higher gas pressure.

Vacuum interrupters in contrast, scale very differently. Making the already very good vacuum even better does not result in a higher dielectric strength. Increasing the contact gap in open position beyond the some 10 mm of the existing medium voltage vacuum interrupters only marginally improves the withstand level. Hence, increasing the voltage ratings of vacuum interrupters beyond the 72.5 kV voltage class is complicated, as it requires several vacuum bottles in series.

Vacuum circuit breakers for 145 kV with two interrupters in series have for long been available from a few (mostly Japanese) manufacturers. Although 145 kV vacuum breakers are widely used in Japan and a few other markets, single pressure SF$_6$ circuit breakers with only one arcing chamber per pole have been the dominating solution for this rating. For higher voltage levels, SF$_6$-based technology stands out even more favourable, mainly because fewer interrupters in series are required. SF$_6$ circuit breakers with only one arcing chamber are available up to 300 kV. Making a 420 kV vacuum circuit breaker would require around six interrupters in series in each pole, with the contacts operating almost perfectly synchronous. The operating mechanism becomes technically complicated and thus expensive. Moreover, conducting away heat generated in the contacts by resistive losses is far easier when the arcing chamber is filled with SF$_6$ gas.

In conclusion, it is reasonable to assume that the application areas of vacuum technology may expand somewhat, but it is hard to envision that vacuum will replace SF$_6$ for circuit breakers at the high end transmission system voltages.

References

1. Gentsch D, Goettlich S, Wember M, Lawall A, Anger N, Taylor E (2014) Interruption performance at frequency 50 or 60 Hz for generator breaker equipped with vacuum interrupters. In: Proceedings of XXVI international symposium on discharges and electrical insulation in vacuum. Mumbai, India, pp 429–432
2. Schellekens H, Gaudart G (2007) Compact high voltage vacuum circuit breakers, a feasibility study. IEEE Trans Dielectr Electr Insul 14(3):613–619
3. Liu X, Lai Z, Cao Y, Liu X, Zou J (2011) Research on insulation characteristics of multi-break in series extra high voltage vacuum circuit breaker. In: Proceedings of the 1st international conf. electric power equipment—switching technology. Xian, China, pp. 44–47
4. Is limiter: the world fastest limiting and switching device, ABB product catalogue
5. Schmitt H et al (2012) Application and feasibility of fault current limiters in power systems, Cigre WG A3.23 report no. 497, June 2012
6. Paul W, Chen M (1998) Superconducting control for surge currents. In: IEEE Spectrum, May 1998, pp 49–54
7. Noe M, Hobl A, Tixador P, Martini L, Dutoit B (2012) Conceptual design of a 24 kV, 1 kA resistive superconducting fault current limiter. IEEE Trans Appl Supercond 22(3):5600304
8. Moscrop JW (2013) Experimental analysis of the magnetic flux characteristics of saturated core fault current limiters. IEEE Trans Magn 49(2):874–882
9. Yuan J et al (2015) Performance investigation of a novel permanent magnet biased fault current limiter. IEEE Trans Magn 51(11):1–4

10. Abramovitz A, Smedley KM, De la Rosa F, Moriconi F (2013) Prototyping and testing of 15 kV/1.2 kA saturable core HTS fault current limiter. IEEE Trans Power Delivery 28(3):1271–1279
11. Abramovitz A, Smedley KM (2012) Survey of solid-state fault current limiters. IEEE Trans Power Electron 27(6):2770–2782
12. Dordizadeh P, Gharghabi P, Niayesh K (2011) Dynamic analysis of a fast acting circuit breaker drive mechanism. J Korean Phys Soc 59(6):3547–3554
13. Callavik M, Blomberg A, Häfner J, Jacobson B (2012) The hybrid HVDC breaker: an innovation breakthrough enabling reliable HVDC grids. In: ABB grid systems, Technical paper, Nov 2012
14. Preve C et al (2016) Validation method and comparison of SF_6 alternative gases. D1-205, CIGRE
15. Seeger M (2016) Survey on SF_6 alternative gases for switching. CIGRE SC A3, Paris
16. Saxegaard M et al (2015) Dielectric properties of gases suitable for secondary medium voltage switchgear. 0926, CIRED
17. Hyrenbach M et al (2015) Alternative gas insulation in medium-voltage switchgear. 0587, CIRED
18. Kieffel Y et al (2016) Green gas to replace SF_6 in electrical grids. IEEE Power Energ Mag 14 (2):32–39

Index

Printed in the United States
By Bookmasters